Devil in the Mountain

Devil in the Mountain

A SEARCH FOR THE
ORIGIN OF THE ANDES

✣

SIMON LAMB

Illustrations by Gary Hincks

PRINCETON UNIVERSITY PRESS
PRINCETON AND OXFORD

Published by Princeton University Press
41 William Street, Princeton
New Jersey 08540
In the United Kingdom: Princeton University Press
3 Market Place, Woodstock
Oxfordshire, OX20 1SY

Library of Congress Cataloging-in-Publication Data

Lamb, Simon.
Devil in the mountain: a search for the origin of the Andes / Simon Lamb.
p. cm.
Includes bibliographical references and index.
ISBN: 0-691-11596-6 (acid-free paper)
1. Geology–Andes. 2. Andes. I. Title.

QE230.L36 2004
558–dc22 2003064124
British Library Cataloging-in-Publication Data is available

This book has been composed in Sabon
Printed on acid-free paper. ∞
pup.princeton.edu

Printed in the United States of America

10 9 8 7 6 5 4 3 2 1

✣ *Contents* ✣

Map of South America showing the main features of the Andes.

⁘ *Preface* ⁘

EVERY YEAR, almost without fail, somebody will ask me if I have any exciting travel plans. I normally slide away from this question with an answer along the lines: "Not really, though I might be going to Bolivia again." But as the years have passed by, I have begun to realize that more and more of my life has been caught up in these Bolivian trips, and I have a story to tell. And I had better tell this story before I forget the details.

I am a geologist, and for more than a decade my work has involved spending several months each year in the Bolivian high Andes, trying to understand the origin of these mountains. You may find this a rather mysterious activity unless you are a geologist yourself. This book is an attempt to remove the mystery, revealing at the same time the deeper scientific understanding of the creation of mountain ranges. And so the structure of the book is based on the idea that the personal story of a scientist—in this case, my story—as he struggles with a scientific problem will light up the subject. I want to take you, the reader, through many of the same experiences, in much the same way that I would show a colleague around in Bolivia, trying to convince him or her that my ideas are right.

I realize that by telling the scientific story from such a personal viewpoint I am in danger of giving the impression that only my observations or ideas matter. This is certainly not my intention, and I am very much aware that my work has been influenced and guided by many other scientists. The Andes are a very big mountain range, extending for over five thousand kilometers in length. The Bolivian Andes alone are more than enough for any individual scientist. Without the combined endeavors of a community of geologists, working over many years, our knowledge of these mountains would be sparse indeed.

It is inevitable, though, that some of this scientific work has been duplicated and similar conclusions have been arrived at independently—in fact, this is good for science because it provides internal confirmation of many scientific ideas. It is no excuse for not giving scientific credit where it is due—nothing is more important to a scientist than proper recognition of his or her work, and I have tried to be as honest and fair as possible, acknowledging my fellow geologists' work in the text.

I have never lost the thrill of making my own observations, or, as a scientist will say, collecting my own data. So by telling my personal story I hope to convey this excitement.

I owe a debt of gratitude to many people. Without the enthusiasm, support, and inspiration of John Dewey, the Andean project would never have come into existence. The research was possible only because of the sustained financial support from the Austrian Academy of Sciences, British Petroleum, European Union, Exxon, Natural Environment Research Council, and the Royal Society of London. Over the years, Catrin Ellis-Jones, Christina Guldi, Anne Grunow, Christy Hanna, Martin Shepley, Eduardo Soria-Escalante, and Marc de Urreiztieta have provided companionship and invaluable help in the field, putting up with both myself and the harsh and difficult environment. The Maxwell family and Molly Relling generously provided both hospitality and somewhere to live during the early stages of writing this book, while I was on sabbatical leave in New Zealand. Evelyn Jenkyns, Grant Heiken, Angharad Hills, Venice Lamb, Ted Nield, George Philander, Tony Watts, and Maarten de Wit courageously and generously read various versions of the manuscript, and their encouragement and advice have enormously improved the book.

My parents have been a vital source of support during my research career, always ready to come to my aid in times of trouble. My father's experience and calm advice have solved many crises that threatened to sink the whole research enterprise. Gary Hincks created the wonderful illustrations, which, I believe, add so much to the book. And without the staunch support and wise advice of my editor, Joe Wisnovsky, this book would never have seen the light of day. Finally, I owe so much to my partner in life, Felicity Maxwell, who has helped me in so many important ways, not least because her careful and patient reading resulted in a better book.

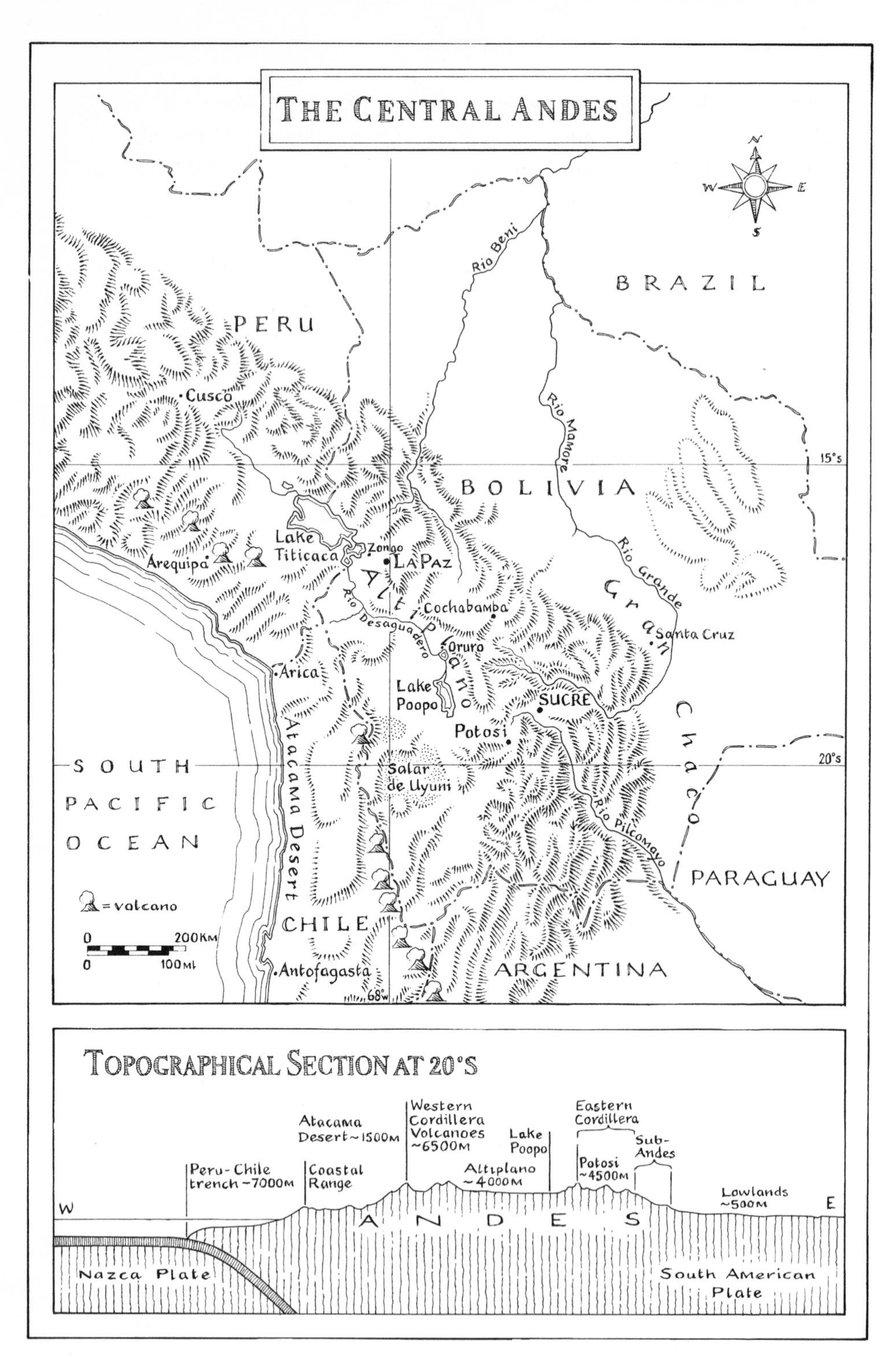

Map of the Central Andes, showing the main features of the Andes in Bolivia, southern Peru, northern Chile, and northernmost Argentina, with the main rivers, lakes, and volcanoes.

✣ *Prologue* ✣

THE LAB FLIGHT TO LA PAZ, Bolivia, had left Miami at just after midnight and was now traversing the interior of South America, heading southward. In the early dawn light, the Amazon jungle, far below, was no more than a grainy texture. Eventually, as the spin of the Earth brought the Sun above the horizon, the forest canopy turned dark green, and sunlight soon pierced the windows of the plane, skewering the cabin with bright bars. On the western margin of the lightening sky, a jagged dark edge, like the teeth of an upturned saw, seemed to float above a long band of cloud.

During the next hour, these teeth—viewed from the aircraft—grew progressively larger, finally revealing themselves as a range of rocky peaks, capped by deeply crevassed patches of cobalt-blue ice. The flight was passing over the high spine of the Cordillera Real—the Royal Range—in the Bolivian Andes. These are the highest of a succession of ranges that extend westward from the Amazon jungle, steadily increasing in height; dense growth covers the lower ranges, in the foothills, but the mountains of the Cordillera Real are barren, elevated thousands of meters above the natural tree line.

PART ONE

✣

✣ CHAPTER ONE ✣

Devil in the Mountain

Devil, ***n*****.** Superhuman malignant being; fighting spirit, energy or dash in attack; mischievously energetic, clever or self-willed person; literary hack doing what his employer takes the credit and pay for. (*OED*)*

A Simple Question

In Bolivia, the local silver miners consider the entrance tunnel of a mine, so laboriously hewn into solid rock, as the gateway into a mysterious realm ruled by earth spirits. The miners are burglars in this underground world, stealing its riches. They are at the mercy of the wicked Tio — the devil in the mountain — who decides the fate of each mine. The miners place a grotesque statue of the devil near the mine entrance to constantly remind themselves of his presence. Each time that they pass this statue, they adorn its head with small presents of cigarettes, money, coca leaves, or sticks of dynamite, hoping that these will keep the devil happy, ensuring the success and safety of their mine.

To a certain breed of scientist — geologists — the devil in the mountain can be interpreted another way. It is geology, the sum total of things that make up the bedrock of our world — the stuff we live on. This bedrock contains a story of events going back to the remote past of our planet, a story that will show, among many other things, why one mountain contains a rich seam of silver while another is just barren rock. In the past two hundred years, scientists have completely lost their fear of the devil in the mountain, developing techniques to chronicle the sequence of events that make up the history of the Earth.

*The chapter opening epigraph definitions throughout the book are paraphrased from the Oxford English Dictionary.

Today, geologists are not content to just reveal the planet's history. Like historians who delve into long past human conflicts, they also want to know why these events happened, searching for the underlying fundamental laws that govern them. There are many lines of geological investigation, and different geologists have pursued answers to their own particular questions. Some geologists—or rather geophysicists—have searched for the fundamental engine that drives the restless Earth, exploring the deep interior of the planet. Others have tried to understand the causes of past changes in the Earth's climate, probing the behavior of the planet's atmosphere and oceans. Yet others—paleontologists—have puzzled over the factors that have controlled the course of the evolution of life, triggering great explosions of life or its demise during mass extinctions. Most geologists are content with tackling one small part of the overarching themes, exorcizing their own particular "devils" and thereby fitting a few extra pieces into the overall jigsaw puzzle of scientific understanding. They have their own fascinating tales to tell of their particular scientific investigations, and the combined effect of all these endeavors has slowly begun to reveal a much bigger picture of how our planet works.

This book describes how I, as a geologist, tried to solve a scientific problem that has long fascinated me. It is the story of the search for the answer to a simple question: why are there high mountain ranges on the face of the Earth, in the continents? Most of the world lies far beneath the sea, and dry land is on average only a few hundred meters above sea level. So, high mountain ranges, reaching more than 2,000 meters (6,500 feet) above sea level, form only a tiny fraction of the Earth's surface. Despite this, they are hardly features of the planet that one can miss. The great mountain chains seem to tower above the rest of the world, following distinct zones that snake across Europe and Asia, from the Pyrenees and along the Alps to the Himalayas and Tibet, or along the western edge of the Americas from Alaska down to Patagonia. From the perspective of our busy lives, on the plains or deep down in the valley bottoms, they can appear vast and forbidding. And from their summits our human world looks decidedly small and unimportant.

Exploring mountainous regions has always been difficult. For this reason, I suppose, I have always been intrigued by these parts of the Earth, where a high ridge of rock can wall off a valley from its neighbors, creating an isolated world. But what forces or processes in the Earth govern their location, overall shape, and height? Some of the greatest scientists, going back at least as far as the ancient Greeks, have tried to answer this question. And it is not a dry and dusty academic question, but one that is central to understanding many important aspects of our planet, including our own evolution. With the specter of global warming looming on the horizon, scientists have been looking ever more closely at the planet's climate. It has become clear that mountain ranges can exert an enormous influence on this climate, determining not only its temperature but also the distribution of rainfall. It follows that the rise of mountain ranges in the past, or their subsequent demise, has driven profound changes in the climate. Biologists are beginning to see that these changes are one of the ways that the planet steers the course of life's evolution.

By chance, my own search for an answer has led me to the high Andes of Bolivia, where over the past fourteen years, on many expeditions, I have tried to chronicle the geological history of these mountains. And so the book contains another story, the day-to-day story of my journeys in this extraordinary part of the world.

A Conversation with Rocks

Anybody listening to two geologists talking would be struck by one thing above all—the sheer range in scale of the subject matter. Geologists habitually think about processes that occur on time scales that are far outside the human experience, taking place over thousands to billions of years. They will observe phenomena that are on a scale of microns (millionths of a meter) to thousands of kilometers. And they will try to weave all these time and size scales into an overall understanding of the way the planet works.

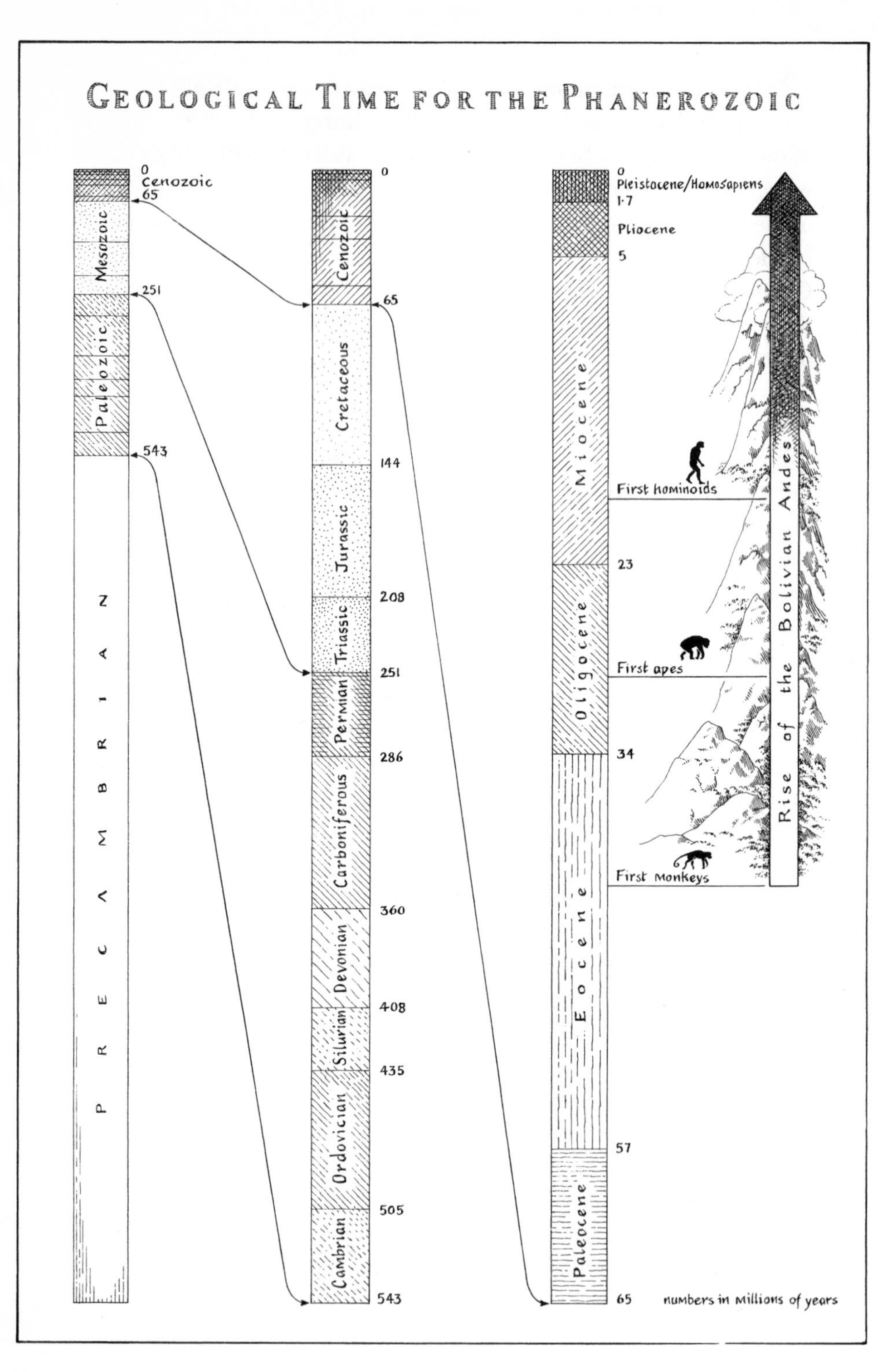

Geological time scale.

At some point in any geological investigation, a geologist will venture out of the laboratory into the real world, or as a geologist will say, into the field. A considerable part of a geologist's training is concerned with the basic craft of fieldwork—something that some people are naturally better at than others. Good field geologists have an uncanny ability to make the right observations at the right time, to sort out the simplicity of the natural world from all its complexity. Being among them is a rare chance to hear nature speaking clearly. The apparent ease with which they make their observations is often deceptive—on one's own again, the world looks as cluttered and messy as ever.

Geologists go where the rocks are. The rocks are not always in easy places. And, unlike tourists, who usually accept much of what they are told on their fleeting visit, geologists continually observe and ask questions, getting under the skin of the landscape they study. Ideally, a geologist would like to be able to flit from one place to another with an all-seeing eye, effortlessly taking in the key observations. Reality is very different. Much of your time is spent just getting around. And often you have no clear idea of where to look, or even, sometimes, what you are looking for. But somehow, if you persevere, the observations start to create patterns or ideas in the mind, and these grow and grow until you seem to be involved in a curious dialogue: the rocks speak, you reply with a question, and they answer back.

Birth of a Project

Look at any map of the world and your eye will be caught by the long, sinuous mountain chain on the western margin of South America. These are the Andes, forming the second largest mountainous region on Earth after the great ranges of Asia. The name conjures up images of high mountain countries, once ruled by the Inca empire and conquered by the Spanish conquistadors in the sixteenth century. The Spanish subsequently extracted a vast mineral wealth of silver and gold from the mountains—a wealth that financed the

Spanish empire. Today, the region is alive with geological activity, rocked by large earthquakes and explosive volcanic eruptions, a region rich in pickings for a geologist.

The Andean project was the idea of John Dewey, professor of geology at Oxford University. In the early 1970s Dewey was one of the few geologists who saw how to use the new, exciting theory of plate tectonics to make sense of a vast range of geological ideas and observations. He had shown that the rocks in the continents contained a record of plate tectonics going back hundreds of millions of years. And he could explain the variety of rocks found in many mountains in terms of the opening and closing of long vanished oceans.

John Dewey believed that the Andes were the place to solve many fundamental problems about the Earth, such as the reasons for earthquakes and volcanoes, the forces that drive the movements of its surface, and, ultimately, the origin of the continents themselves. Though the Andes were deeply embedded in the geologist's jargon—a common type of volcanic rock, found in many parts of the world, is called andesite after the Andes—they were somehow being neglected by geologists. The eye of the geological community was turned toward the Himalayas and Tibet. Yet, as we shall see, many geological clues to the way the Earth works were first found in the Andes.

The late 1980s was a good time to think about working in South America because a number of big, multinational oil companies had decided to invest heavily in oil exploration there, both in the Andes and along its eastern margin. One of these companies agreed to pay for a five-year research project that would provide background information for their own geologists working in the region. They left the details deliberately vague. The idea was to let the project have the room to develop in its own way. Whether the company was right, time will tell, though without doubt this unusual freedom gave rise to a very unusual research project.

In 1989 I was at a loose end. I had recently returned from New Zealand, where I had been studying the effects of earthquakes and fault lines along the boundary between two great tectonic plates.

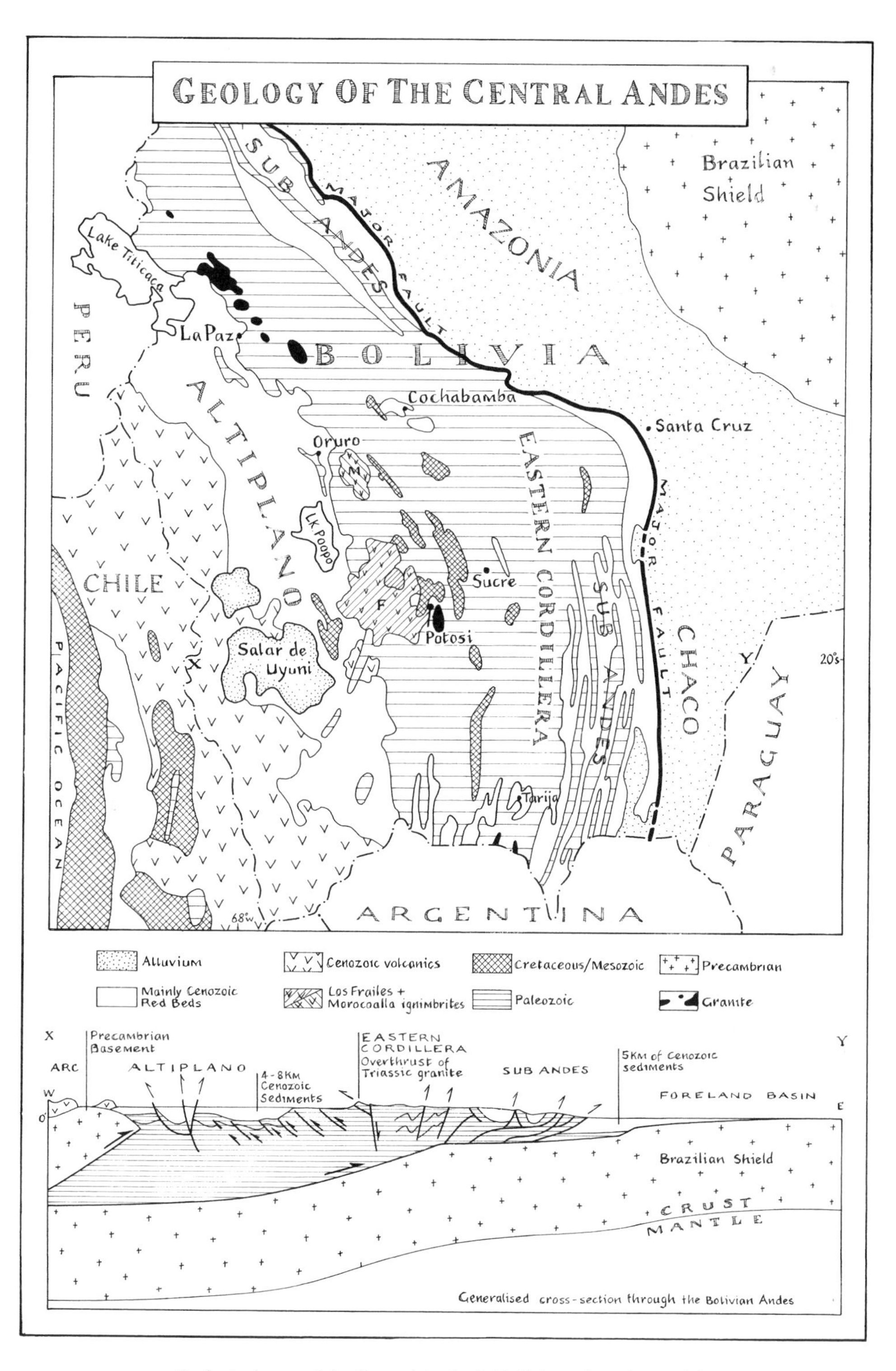

Geological map of the Central Andes in Bolivia and northern Chile.

I had just presented the results of some of this work at a conference in Greece. Here, I met up again with Philip England, who had been recently appointed as a geophysicist at Oxford University. Philip passed on my name to John Dewey, who was on the lookout for somebody to run his South American project. I had no idea that the job would occupy more than a decade of my life. During this period, Dewey gave me complete freedom to travel where I wanted and develop my own research.

Destination Bolivia

Our choice of Bolivia as the focus for the project started with a telephone call. I rang up the Foreign Office in London for advice about working in South America and was put through to a man on the South American desk. I told him that I was running a scientific research project in the Andes. Did the Foreign Office have any particular advice about undertaking this sort of work in South America? His reply made me feel rather gloomy:

"Well, I wouldn't go to the northern part of South America—it is far too dangerous with the various drug wars. Peru is out because of the Sendero Luminoso guerrillas—the Peruvian government have lost control of most of the country. There have been a number of military rebellions in Argentina—the country is still very unstable after the Falklands disaster and they don't like the English. Chile is a bit tricky at the moment because there is strong popular feeling against General Pinochet and it is unclear whether he is going to give up power voluntarily."

He paused for a while, then said rather doubtfully:

"Well, there is Bolivia."

I knew that Bolivia was very poor and that it was supposed to have been the scene, near San Vicente, of many of Butch Cassidy and the Sundance Kid's more daring robberies, stealing money from the rich mines. Also, I had always thought that Bolivia was the most dangerous and difficult South American country of all.

"Bolivia?" I echoed.

"Yes"—he sounded much more confident now—"do yourself a favor and go to Bolivia. I can't see any real problems there at the moment. The currency is fairly stable and they have a democratic government."

This final comment has to be seen in the light of the fact (as I subsequently discovered) that in the roughly two hundred years of the Republic of Bolivia, there has been on average a change of president every six months—usually as a result of a military coup. Also, in the late 1980s Bolivia had just emerged from an unusually severe bout of hyperinflation, with an annual inflation rate of 24,000 percent. When I first arrived in La Paz in 1989, the smallest note denominations were one million and ten million pesos—these had been optimistically relabeled as one and ten Bolivianos. A standard scam was to palm off on unsuspecting tourists million-peso notes as change instead of ten million pesos—you had to be quick thinking to notice whether there were six or seven zeros after the one.

Curiously enough, it made good scientific sense to base our project in Bolivia. The mountain ranges of the Andes are highest and widest there, and there is an intriguing swing in their general orientation, following the curve of the coastline. Since first looking at maps of South America, I had a feeling—which I had never quite thought through properly—that the key to understanding the origin of the Andes lay in the high peaks of Bolivia. For this reason, after my initial reaction, I was very receptive to the idea of working there.

I probed the man on the South American desk a bit more to reassure myself that he was serious about his suggestion, then hung up with a feeling of excitement. I was beginning to feel that what at first had seemed to be an academic exercise was developing into an adventure. But, if you really want to know how this adventure unfolded, I need to prepare you for it, trying to put you into my shoes, as it were, when I first embarked on my research. The best way to do this, I think, is to begin telling a story stretching back over two thousand years—a story that I find fascinating in its own right—of how geologists have struggled to make sense of mountains. Their discoveries formed the basis of the stock of ideas and preconceptions that I set out with—some of these I soon discarded, but others I came to value more and more.

In everything that follows, I have tried to be as self-explanatory as possible. However, it is difficult to discuss many geological phenomena without making use of some basic ideas or jargon. Though I have gone to great lengths to unpack these wherever I first use or introduce them, you may still find it helpful to remind yourself of the meanings of particular terms, especially when they reappear later on, by dipping into the selected glossary at the back of the book.

✣ CHAPTER TWO ✣

A Mountain of a Problem

Mountain, ***n.*** Large natural elevation of earth's surface, large or high and steep hill, esp. one over 1000 ft. high.

Problem, ***n.*** Thing hard to understand. (*OED*)

IF YOU DRIVE NORTH of the fishing port of Ullapool, far up on the northwest corner of Scotland, you will eventually come to Inchnadamph on the shores of Loch Assynt. You would be forgiven for missing Inchnadamph altogether as it consists of no more than a hotel and a handful of scattered houses. Even less easy to find is a small cairn perched on top of a low hill overlooking the loch. A bronze plaque on the side of the cairn preserves the names of two British geologists, Benjamin Peach and John Horne, who were sent in 1883 by the British Geological Survey to make a geological map of the rocks in this part of Scotland. Peach and Horne walked or rode over virtually every inch of the remote and rugged moorland and hillsides, covering thousands of squares kilometers, recording and measuring the rocks. They eventually produced in 1892 what many geologists today consider to be the pinnacle of geological achievement, quite simply, the finest geological map ever made.

I am very familiar with Peach and Horne's map because I take a class of undergraduate geology students to this part of Scotland every year. We often stay at the Inchnadamph Hotel. Here, in the lobby, is a photograph of Peach dressed in the tweed outdoor clothes of the time, wearing a deer stalking hat and leaning on a stout walking stick, resting after a day showing some of Europe's most eminent geologists his discoveries. And in the hotel's visitor book, in a beautiful flowing script, are Peach's and Horne's signatures. Countless times, after a hard day teaching the students about geological map making, I have escaped the noisy and crowded hotel and made my way to the Peach and Horne memorial. There, the landscape seems peaceful in the evening light, disturbed only by the flutter of bats chasing midges chasing myself. But the rocks in the rugged hillsides contain an extraordinary story of upheavals of the Earth's surface on a scale not believed before Peach and Horne started making their map.

Peach and Horne's arrival in the Scottish highlands was an important moment—a moment I will return to—in a much longer scientific endeavor to understand the origin of the Earth's surface. As with many of the great voyages of discovery, an assorted band of philosophers, theologians, scientists, and geologists (and some sailors!) set off from home into unchartered seas, trying to come to grips with a strange world in which they could no longer rely on familiar assumptions. Here, mountains became terrifying phenomena created out of the fundamental instability of the planet's bedrock. This chapter is a record of the early stages of this voyage, pioneering the way for our own work in the Andes.

Seashells and Floods

Myths, I believe, often contain a kernel of truth, even if this truth is not understood by those most familiar with the stories. A recurring theme in the myths of many early civilizations is the notion that the Earth's surface, early in its history, was sculpted by a violent paroxysm—a catastrophe that transformed an old world, designed by God, into one with its present varied landscape, punctuated with mountains, undulating lowlands, and deep oceans. Violent events—earthquakes, volcanic eruptions, storms—are part and parcel of our world. So, perhaps this harking back to some early primeval catastrophe is no more than the memory in any society of the last big event, which, from a human point of view, may seem overwhelmingly large.

Flooding, especially for peoples who live in a delta region, as did the earliest Middle Eastern civilizations, is a catastrophe. And if the known world does not extend far, it is easy to see how the scale of such an inundation, stretching as far as the eye can see over wide river flats, could be exaggerated. It took only the sharp eye of a Greek geographer such as Herodotus, who noticed in the 400s B.C. that seashells could be found far from the sea in Egypt "upon the mountains of it," to suggest that there was actually hard evidence for ancient floods of huge proportions. One such flood, enshrined in the Genesis story of Noah's ark, has exerted an enormous hold over European imaginations. Another version of this story was written down by the Assyrian scribes in the 600s B.C. and preserved on clay tablets in the library of Ashurbanipal:

> The people of the land watched, bewildered and quiet, as Adad turned all that had been light into darkness. The powerful south wind blew at his side, uniting the hurricane, the tornado, and the thunderstorm. It blew for a full day, increasing speed as it traveled, and shattered the land like a clay pot .... the storm wind raged furiously over land like a battle. It brought forth a flood that buried the mountains and shrouded the people .... Its attacks ravaged the earth, killing all living creatures and crushing whatever else remained.

To many commentators, the biblical flood was both unique and turbulent enough to stir up the entire surface of the Earth, reforming it into a new lumpy terrain. In this way, the perfection of the antediluvian world was forever destroyed by a God exacting retribution for human evilness. This view of the past seems to have remained unquestioned for nearly a millennium, except for a few lone voices, such as that extraordinarily gifted painter, sculptor, musician, inventor, engineer, and scientist, Leonardo Da Vinci, crying out in the philosophical wilderness. Leonardo clearly believed the biblical flood story literally, but he argued, around 1500, that seashells were too heavy to be washed up onto mountain tops, and there was not enough time during the forty days of the flood—"as he has said who kept a record of that time"—for them to swim there. In other words, he doubted whether the flood waters would have the power to reshape our world.

By the end of the seventeenth century, philosophers seem to have dealt to their own satisfaction with Leonardo's objections, lending further credibility to the reality of the biblical flood, with all its presumed consequences. Nicolaus Steno, a Danish physician and mathematician, proposed that the flood waters came from deep inside the Earth. So, rather than stirring up the Earth's surface from above, the rising waters undermined it from below by leaving behind void spaces into which the overlying rock layers quickly collapsed. This idea was summarized by Steno's contemporary, Gottfried Leibnitz, the great German mathematician and rival of Newton, in the following terms: "Enormous bubbles .... have burst, here and there, so that certain portions subsided to form the trough of the valleys, whereas others, more solid, have remained upright like columns and, for that reason, constituted mountains." Leibnitz goes on to say that "the water ... would be forced from the depths of the abyss across the wreckage, and, joining that which was naturally flowing from high places, would give rise to vast inundations which would leave abundant sediment at divers points." The sediment was the accumulation of rock fragments, plucked from the bedrock by the turbulent flow of water. Steno saw in this a way for the Earth to acquire a new rocky veneer, laid down in horizontal layers, on the floor of the depressions between mountains. In effect, he was acknowledging the possibility of some reshaping of the Earth's surface after the flood.

It is an interesting comment on the intellectual environment in which Steno was working that he thought it necessary to devote considerable space in his writings to disproving the notion that mountains were like vegetables or animals and grew organically. J.R.R. Tolkien—steeped in the meaning of Anglo-Saxon language and mythology—must have been familiar with such ancient beliefs when he wrote his fantasy story *The Hobbit*. Early in the book, Bilbo Baggins and the band of dwarves are captured by three trolls called William, Bert, and Tom. The trolls cannot agree on the best way of cooking the dwarves and end up arguing all night:

> Just at that moment the light came over the hill, and there was a mighty twitter in the branches. William never spoke for he stood turned to stone as he stooped; and Bert and Tom were stuck like rocks as they looked at him. And there they stand to this day, all alone, unless

the birds perch on them; for the trolls, as you probably know, must be underground before dawn, or they go back to the stuff of the mountains they are made of, and never move again. That is what happened to Bert and Tom and William.

Steno forcefully rejected this idea out of hand, writing sternly, "There is no growing of mountains .... the rocks of mountains have nothing in common with the bones of animals .... since they agree in neither matter nor manner of production." It is easy to laugh now, but I actually think, despite Steno's best intentions, that there is a grain of scientific sense in the idea of living or growing mountains, though obviously not in the way that Steno was arguing against, or Tolkien described; this is something I hope to make clearer when I describe my own work in the Andes.

Steno also had to counter the objections of theologians, who now argued that the seashells preserved in the rocks were not fossils at all, but sports of nature, randomly formed by the creator to confuse mankind. As such, they had no significance, other than revealing the creator's sense of humor. Finally, right at the end of the seventeenth century, the Reverend John Woodward—who endowed the first geology professorship at Cambridge University so that someone would take an interest in his mineral collection—realized that the existence of fossils could work, after all, in favor of orthodox theological thinking. He reasoned that a rational God would not play tricks on us. Fossils were, indeed, just what they seemed to be—the remains of seashells or other creatures. And fossilized seashells made perfect sense to Woodward because they were some of the remains of the creatures destroyed during Noah's flood!

The real scientific advance in all this must surely be the development of the idea that there is a record on the Earth of what the planet was like in its remote past, and its surface might have changed out of all recognition. But, beyond this, very little progress in geological understanding had been made, and even if Noah's flood was capable of transforming the landscape, it did not explain how landforms, such as those mountains "that were covered to a depth of more than 20 feet" by the rising flood waters, were created in the first place. In 1650 Archbishop Ussher had famously calculated from the long list of Old Testament genealogies that the Earth was created in 4004 B.C. This date severely limited the scope of possible mechanisms for mountain building or, indeed, any other geological activity on the planet.

The Abyss of Time

It seems that it was not until the mid-eighteenth century that there was what I would call a major scientific breakthrough. The French naturalist, George Leclerc, Comte de Buffon, drastically changed the time scale for the Earth's history, adding more time. And more time brought with it new possibilities for mountain building. Buffon thought that the planet must have once been hot like the Sun, forming an incandescent body of molten rock. His experiments with white-hot balls of metal suggested to him that an originally molten body, the size of the Earth, would take tens of thousands of years, at the very least, to cool, shrinking at the same time. He calculated that it would be sufficiently cold roughly thirty-five thousand years ago for the oceans to condense, covering the planet with a layer of water. Here, life could thrive. By then the shrinking would have caused the outer skin of the planet to wrinkle into bumps—undersea mountain ranges—and depressions. Buffon neatly reversed Steno's idea by proposing that the water gradually withdrew into the cracked interior of the cooling planet, leaving behind oceans in the depressions and seashells stranded on the emergent mountains.

The Scot, James Hutton, went much further than Buffon, proposing a history for the Earth and mountain building that was far longer than even tens of thousands of years—in fact, in Hutton's mind, it was virtually limitless. Hutton believed that his ideas were proved by some key geological observations. In 1784 he had found in Glen Tilt, in the Scottish highlands, a body of pink granite. His companions describe how he immediately began shouting and jumping up and down. Eventually, he calmed down enough to explain this spontaneous burst of emotion—he could, at last, prove that granite was once molten and had intruded into the surrounding rocks as a liquid. There were sure signs of this where veins of granite extended from the main body, sometimes wrapping round blocks of the host rock. In effect, he had killed off the widely held idea that granite formed by precipitating from sea water. But why was there molten rock here in the first place? Hutton suggested an answer that was inspired by what he had seen in the cliffs at Siccar Point, southeast of Edinburgh. Here, gently tilted layers of sediment rest directly on top of older contorted rocks layers, forming a contact now known as Hutton's Unconformity. The older rocks have been twisted to such an extent that their once horizontal layering is turned up vertically.

To Hutton, the rocks could not speak more clearly. He saw the surface of the Earth as being in a constant state of up and down motion, forming a sort of natural seesaw that operated endlessly over the vast periods of geological or deep time, as some people call it. Rivers relentlessly flow off the highlands, carrying away detritus down to the sea, where it accumulates in horizontal layers. This way the highlands are eventually worn down. But in their destruction, according to Hutton, lies the seeds of new highlands, because as the detritus is buried on the sea floor, it is heated up in the Earth's interior. This, Hutton thought, would ultimately cause the pile of sediment to expand again, locally even melting to form the granite he had observed in Glen Tilt. The expansion would lift the rocks out of the sea, creating new highlands and also twisting and contorting the rock layers so that they stood on end. But there was time enough for these new highlands to both cool and be worn away by rivers, eventually subsiding back into the sea. And so the whole seesaw motion would start again, as more detritus accumulated above them, to be, in time, tilted and raised again out of the sea to where Hutton found them. One of Hutton's students, John Playfair, on being shown by the master the evidence for this extraordinary sequence of events at Siccar Point, wrote: "On us who saw these phenomena for the first time, the impression made will not be easily forgotten.... What clearer evidence could we have had of the different formations of these rocks, and of the long interval which separated their formation. . . . Revolutions still more remote appeared in the distance of this extraordinary perspective. The mind seemed to grow giddy by looking so far into the abyss of time."

Playfair's moment of giddiness marks an important step forward in our understanding of the origin of the Earth's surface with its high mountain ranges. Whatever else one might think about Hutton's seesaw vision of the Earth's surface—and there certainly are elements that are not far off what a modern geologist might think—one can only admire his attempts to learn the language of the rocks. Like any new language, there were many words whose meanings were obscure, so Hutton, and the geologists he had inspired, were forced to bridge these gaps with preconceptions of their own. They desperately needed a reliable dictionary. But it was not until the early 1830s that a really useful reference book was available. By then, Charles Lyell had published his *Principles of Geology*, a book that would deeply influence geologists for many years to come. Lyell shared Hutton's belief of an almost limitless time scale for geological activity. But, crucially, he only considered processes that were going on today at the Earth's surface as an explanation for the story in the rocks, an approach sometimes summed up in the catch phrase "the present is the key to the past."

Lyell believed that the history of the Earth consisted of a succession of innumerable small changes in the environment, operating over the vast vistas of time, rather than sudden global cataclysms—the record in the rocks tended to telescope together this more gradual evolution, giving the appearance of a catastrophe. These concepts planted a seed in the mind of the young Charles Darwin, who had been one of the first to read Lyell's great work. This seed was to take root and sprout into a veritable tree of an idea, explaining mountains in terms of what is happening today on the surface of the planet. And I find it particularly satisfying that Darwin's inspiration was provided by the Andes.

A Great Earthquake

Charles Darwin visited the western side of South America in 1835 on the round-the-world voyage of HMS *Beagle*. On the night of January 19, while the ship was anchored off the coast of central Chile, a "great glare of red light" was observed toward the land. Subsequent investigation revealed that the party had witnessed the eruption of incandescent molten rock from Corcovado volcano, which forms part of a vast chain of volcanoes that lie along the high spine of the Andes. Darwin received reports that several other volcanoes, hundreds of kilometers farther north, had also erupted on the same night.

Almost exactly a month later, on February 20, Darwin felt the effects of an exceptionally strong earthquake. He was violently rocked for nearly two minutes while on a shore visit near the coastal town of Valdivia. The full force of this earthquake became clear when the *Beagle* sailed north to the port of Concepcion. The harbor master immediately gave them the terrible news that they had arrived at the scene of a disaster—not a single house in Concepcion remained standing, and seventy outlying villages were also totally destroyed. Huge waves had wreaked havoc along the coast. More worrying to the local people, it appeared that large parts of the coastline had been permanently raised up nearly three meters out of the sea during the few minutes of the earthquake; fishermen reported seeing rocks sticking out of the sea that had not been there before the earthquake.

Darwin realized that he had witnessed in Chile a powerful agent of change within the Earth. The significance of this started to emerge when he explored more of the country. He found thick banks of seashells resting on cliff tops, up to 450 meters above the present sea level. A journey much further inland, across the high Andean mountain ranges, revealed more seashells, this time in tilted layers of rock as much as two thousand meters above sea level. To a naturalist like Darwin, influenced by Lyell and searching for fundamental causes of change in the Earth, it was obvious that the small increment of uplift he had experienced during the recent great earthquake was part of a longer-term process that had raised up the land surface and created the mountains. Darwin went on to speculate that this was connected with the volcanic eruptions he had seen a month before the earthquake. As well as seashells, he had found the remains of large bodies of once molten rock — red granite — in the mountains. He suggested that the force that pushed up the Andes was the upward rise of buoyant liquid rock, as it found its way to the surface from deep inside the Earth. As it rose higher, the liquid rock forced aside the surrounding rock layers, twisting and breaking them. In the process, the Earth literally creaked and shuddered during earthquakes.

Darwin was one of the first scientists to see clearly the creation of mountains in terms of observable processes going on today at the Earth's surface. He had seen for himself in one of the earthquakes that periodically rocked Chile a powerful force to push up the land. And he had made the leap from this to the accumulated uplift during many earthquakes, over geological time, of a high mountain range. Even his link with the movement of molten rock is an idea that geologists still entertain. He wrote in his account of these discoveries: "Daily it is forced home on the mind of a geologist, that nothing, not even the wind that blows, is so unstable as the level of the crust of this Earth."

This extraordinary idea, so contrary to one's basic intuition about the solidity of the underlying bedrock, came from simple observations about the natural world. But new ways of looking at the planet were being developed by a different breed of scientist — astronomers and surveyors who were trying to measure the world. These people had no geological interest in the Earth's crust, yet their work would add a new dimension to the study of mountains, forcing geologists to consider what lay underneath, much deeper in the Earth. And, again, it is the Andes that would be the place where the crucial measurements were first carried out.

A Problem with the Plumb Line

In 1735, exactly a hundred years before Charles Darwin witnessed the great Concepcion earthquake in Chile, Louis XV of France commissioned a team of surveyors to determine the shape of the Earth. This remarkable expedition, led by Charles-Marie de la Condamine, turned up the first clues that large mountain ranges are associated with profound changes in the deeper nature of the Earth. The French surveyors were trying to measure the distance on the surface of the Earth that constitutes one degree of latitude, using a combination of land surveys and the principles of celestial navigation developed by seamen over the centuries as they navigated their ships around the world. Their work lead them to the high Andes in Ecuador where they made observations along a line of longitude between Quito and Cuenca among the Andean mountain peaks, including the towering volcano of Chimborazo.

Modern navigators have largely lost the ability to steer a course by the stars, a technique I need to explain if one is to understand the work of the French expedition. Yet, in many ways, the latest satellite positioning systems are just elaborate versions of the age-old method, in either case using objects in the sky—stars or satellites—as reference points or navigational beacons. These beacons are used for traditional navigation by taking a sighting with a telescope. The aim is to measure the angle—an angle called the altitude—between the line of sight of both the celestial object and the thin ribbon of the horizon, where the sea meets the sky; the horizon defines the local horizontal—the local flat surface of the Earth. In general, navigators fix on a combination of stars, planets, and the Sun—only one point on Earth, defined by both its latitude and longitude, will be consistent with all these measurements.

A surveyor on land—in the Andes, for example—needs to modify slightly the ocean-going navigator's technique. This is because, on land, hills or mountains get in the way of the distant horizon, making it difficult to measure the altitude of a celestial object. What the surveyor does, instead, is use a plumb line. When the weight is allowed to hang free, the line of the string is the local vertical—this is how a builder makes sure that walls stand straight and upright. By definition, the local horizontal will then be at right angles to this vertical—a spirit level will do essentially the same job, giving the horizontal directly. But for surveyors making very accurate measurements, there is a problem with this. They are relying on Newton's law of gravity—the weight at the end of the plumb line is pulled down by the gravitational attraction of the mass of the Earth. But this pull is not necessarily straight downward. Marked bumps or irregularities on the Earth's surface, such as mountains, would be predicted by Newton's law to pull slightly sideways on the weight, tilting or deflecting the plumb line away from the true vertical. This effect, though small, would still need to be taken into account when making precise land observations to determine a star's altitude.

Corrections for the pull of the mountains on a plumb line were too difficult for the French scientists in the Andes of Ecuador, so they decided to just leave them out. Despite this, they found that their celestial observations fitted in remarkably well with other estimates of the length of a degree of latitude, made elsewhere in the world, apparently proving a widely predicted flattening of the Earth at the poles compared to the equator. This result puzzled a junior member of the expedition, Pierre Bouguer, who had expected the pull of the massive Andean mountains to significantly deflect the plumb line away from the true vertical and upset their measurements. It seemed as though the mountains could be ignored after all.

Exactly the same problem was encountered slightly over a hundred years later by British surveyors in India, led by Sir George Everest, when they were fixing points throughout the Indian subcontinent with the greatest precision that was possible at the time. Celestial measurements of position where made, again using a plumb line, on the plains of northern India between Kalianpur and Kaliana. Kaliana lies immediately south of the vast ranges of the Himalayas, including the highest mountain of all, Mt. Everest. Yet, when the surveyors compared celestial observations with accurate land measurements of the relative distance between these two places, they found better agreement if they did not correct for the expected deflection of the plumb line caused by the pull of the mountains. So striking was this inconsistency that it was remarked at the time that the mountains seemed almost hollow, with no mass at all. How could solid mountains be hollow? This was a major paradox.

Icebergs or Risen Bread?

It is often the case in science that a paradox forces scientists to reevaluate their basic assumptions. In 1855 the British Astronomer Royal, Sir George Airy, put his finger on what was wrong with the surveyor's analysis of plumb lines. He realized that the reason why the gravitational attraction of mountains did not cause the predicted large deflection of the plumb line is that there was something underneath the mountains that was counteracting this attraction. In other words, when making these calculations, one should consider not just the rocks at the surface, but also those in the Earth's interior.

Airy imagined a surveyor standing with his plumb line on the lowland plains near a large mountain. Because of gravity, the weight at the end of the plumb line behaves as if it is pulled toward the Earth by strands of gossamer thin elastic thread tugging in all directions. On its own, the sideways pull of the mass of rock in the mountain would be equivalent to tilting or deflecting the plumb line by pulling more sideways on one of the elastic threads connected to the mountain. But what if the rocks beneath a mountain were relatively light weight, with less mass than normal? In other words, what if the mountains had an underlying low-density root? Newton's laws of gravity would predict that such a root would significantly lessen the gravitational pull at depth on the plumb line. In the analogy of the elastic threads, this would be the same as slackening off the tension in the thread connected to the root, so that the deflection of the plumb line eases off again.

If there was a low-density root, then what could it be? In fact, Airy thought that a mountain and its root were one and the same

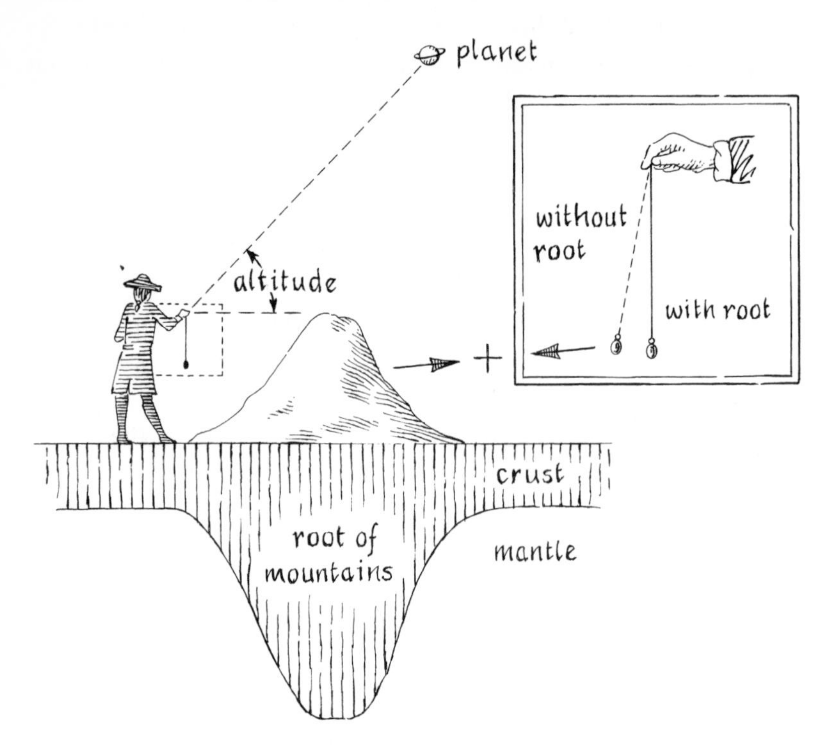

The force of gravity in a mountain of massive rock should deflect a surveyor's plumb line sideways. In fact, the observed deflection is quite small. This discrepancy points to some counteracting action of relatively light rock at depth beneath the mountains.

thing, forming a light crust to the Earth, resting on a higher-density "mantle"; it was just that the crust was much thicker beneath a mountain than in the lowlands. This extra thickness was the root of the mountain. Geologists had already come up with the distinction between the crust and mantle to account for the fact that the common rocks found at the surface—thick accumulations of sediment and granite—are different from the dark lavas erupted from volcanoes such as Mt. Etna in Sicily. It seemed that the volcanoes were tapping some deeper part of the Earth—the mantle—that had a completely different composition from that of the surface crust. Airy proposed that the crust was doing more than just resting on the underlying mantle, it was actually floating in it. Referring to the mantle as the "lava," he described this idea in these terms: "The state of the Earth's crust lying upon the lava may be compared with perfect correctness to the state of a raft of timber floating upon water; in which, if we remark one log whose upper surface floats much higher than the upper surfaces of the others, we are certain that its lower surface lies deeper in the water than the lower surfaces of the others."

Replacing Airy's logs with floating ice, we have an even more graphic analogy for mountains—they are like the tip of an iceberg. The ice pokes above the surface, held up by its deep, underwater icy root. This image immediately implies that the base of the crust is an exaggerated mirror image of the surface topography—where mountains tower above the general surface of the Earth, their roots delve deeper into the mantle.

Unfortunately for Airy, his idea was not the only one to explain the effect of mountains on plumb lines. In 1859 John Pratt proposed an alternative solution to the problem. Pratt was a British mathematician and clergyman who had made the original laborious calculations for the correction of the plumb line in the Indian survey. He may have felt slightly miffed that Airy had effectively highjacked his work before he himself had had the chance to think about its implications. Now, Pratt had his chance. He saw the crust differently from Airy; for Pratt, the crust was rather like bread—there were mountains where it was light and well-risen and lowlands where it was unleavened heavy dough. In this case, the gravitational pull on the plumb line, exerted by the crust beneath the high but lower-density mountains, was counterbalanced by the pull in the opposite direction of the higher-density and equally massive crust beneath the lowlands. In a nutshell, Pratt was proposing that mountains existed because the material that the crust was made of, like the dough and bread, was not uniform. Airy had said nothing about changes in the nature of the crust, merely that its thickness varied.

It turns out that deciding between Airy's and Pratt's explanations, or, for that matter, anybody else's, for the effect of high mountains (or lack of it) on plumb lines is crucial to understanding the origin

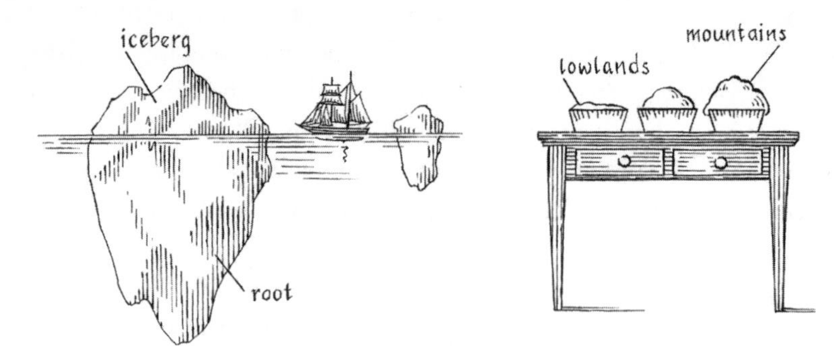

Airy's roots and Pratt's bread: two rival mid-nineteenth-century theories to explain mountains. George Airy thought they were like floating icebergs with roots extending deeper into a fluid substratum. But John Pratt saw mountains as more like risen dough, standing high because they had expanded and were less dense than the rocks beneath the lowlands.

of mountains. But when these ideas were first put forward, nobody had any real sense of how the crust was formed in the first place, and they certainly could not explain why it might have roots extending deep into the mantle, or vary in density. In fact, it was not until 1909, when the Croatian geophysicist Andrija Mohorovičić studied the vibrations triggered by an earthquake in the Pokupsko region of Croatia, that the existence of a crust and mantle was even proved. Mohorovičić found that the earthquake vibrations traveled slower in the crust than in the underlying mantle. This difference in speed marks a change in composition of the rocks. Today, the boundary between the crust and mantle is called the Moho in honor of Mohorovičić—few English-speaking geologists can reasonably hope to pronounce his name correctly.

While surveyors debated about their plumb lines, geologists had, at last, begun to look much more closely at the rocks, amassing evidence of their own that would help to make sense of the crust beneath mountains. This evidence can be found in geological maps. Making these maps is probably the greatest contribution that geologists have made to understanding our planet: the maps are both the inspiration and, critically, a test-bed for almost all geological ideas.

Rock Maps

It is hard to talk about a geological map without making reference to specific types of rocks. So, to begin with, what precisely are rocks? They are the solid substance on which our world rests. However, rocks are not single substances, they are built of a mixture of different minerals in the same way that concrete is a mixture of sand, gravel, and cement. The type of minerals and the way they have formed or come together depend on the origin of the rock. Some rocks were once molten and have crystallized from a liquid—these are referred to as igneous rocks. These rocks are either erupted from a volcano as lava flows or the solidified remains of molten bodies of rock, such as granite, that once lay deep beneath a volcano.

Sedimentary rocks, as even some of the earliest geologists had recognized, are the accumulation of detritus washed down by rivers into the sea, or blown about by the wind, forming sandstones, siltstones, and shale. They can also be the remains of fossilized organisms or chemicals that have precipitated from the water, usually forming limestone or salt. In most cases, the resultant rock owes its existence to the flow of water or wind that moves the sediment around, finally depositing it in roughly horizontal and parallel layers. Geologists often call these layers or strata the bedding. Over time, the nature of the sedimentary layers or bedding may change for a variety of reasons; perhaps the source of the detritus is different as rivers shift or cut down through new rock, or perhaps the nature of the environment in which the sediment is deposited changes. In any case, these changes can give rise to the accumulation of a distinctive sequence of layers over time. As the layers are buried, they are compressed by their own weight and bonded together, rather like cement, by the formation of new crystals of quartz or calcite, eventually turning into solid rock.

Finally, if the igneous or sedimentary rocks are subsequently heated up or subjected to large pressures, they may undergo internal chemical reactions or transformations that can change the nature of the rock. Geologists refer to this sort of "cooking" process as metamorphism—the minerals in the rock are metamorphosed from one form to another, and the delicate preservation of fossils is obliterated. Such transformations are usually the result of deep burial of the rocks, for example, when very thick piles of sediment accumulate. By examining the minerals in the rock it is often possible to calculate the temperature and pressures the rock has been subjected to—certain minerals only form or are stable at particular temperature and pressure conditions.

Armed with the ability to recognize different types of rock, the geological map maker's task is literally to walk out the ground searching for signs of changes in the bedrock. The key to the technique is to locate the bedrock on a detailed topographic map, recording various features such as the type of rock, the orientation of any layering, and the nature of the contact with adjacent rocks. Gradually, a picture builds up on the flat sheet of the map of the three-dimensional nature of the bedrock. Any competent geologist should be able to turn this flat map back into a three-dimensional picture, in much the same way that an architect can read a set of architectural plans to visualize the final building. And like architects, geologists draw profiles or sections through their maps to help them in this visualization. For example, a vertical section through a map of a simple sequence of layered sedimentary rocks would appear much like the view one might have of the sedimentary layers in a cliff, with the youngest layers at the top and the oldest at the bottom.

The skill in geological mapping centers round completing the map from the very incomplete picture of the rocks that is visible at the surface. This is a bit like trying to work out the picture on a jigsaw puzzle when half the pieces are missing. The degree to which "pieces" are missing on the geological jigsaw puzzle is summed up by the amount of rock exposure; anybody who regularly walks through the countryside will know that the rocks are usually hidden beneath a layer of soil. On occasion, I have even resorted to examining loose chips of rock brought up by burrowing rabbits.

A Geological Masterpiece

We now, at last, return to Peach and Horne's map of the rocks in the Assynt region of northwest Scotland. So, what was it that they discovered? They found a distinctive sequence of sedimentary rocks that were originally deposited in shallow coastal waters close to a large continent, such as along the present-day coast of Florida, with similar conditions of current and water depth over a wide region. Layers of white sandstone, full of well-rounded grains of quartz—probably deposited as sandbanks by tidal currents—form the base of the sequence. The shallow sea floor slowly subsided, and layers of shale and limestone settled out on top. We can date when this happened because the rocks contain the distinctive fossils of creatures that lived not long after multicellular animals, with shells or bony hard parts, first appeared on Earth, in the Cambrian and Ordovician periods of geological time—periods named after the pre- Roman tribes that once occupied the parts of Wales where the fossils were first found. And the exact same level or position in the sequence can be pinpointed in rock outcrops over a hundred kilometers apart.

The fossils in the sedimentary rocks are perfectly preserved. This is good evidence that both the fossils and the enclosing rock have never been subjected to high temperatures and pressures. In other words, these are clearly not metamorphic rocks. But Peach and Horne did find metamorphic rocks, making up both the western and eastern parts of their map. Those in the west contain minerals such as garnet and biotite, typical of igneous and sedimentary rocks that have been heated up to high temperatures, over five hundred degrees centigrade, at some later stage in their history. Peach and Horne interpreted them to be the remains of a once deeply buried core to an ancient continent. Later, the action of rivers had stripped away the surface rocks, exposing the metamorphic basement. And even later still, this basement had subsided and was inundated by the sea. It was at this time that the distinctive sequence of sedimentary rocks was deposited on top. The metamorphic rocks in the east could not be so easily explained. These looked completely different. They seemed to be the cooked-up remains of a vast pile of sediment that had been deposited on the floor of an ocean.

The significance of Peach and Horne's work becomes apparent when one tries to visualize a cutaway vertical view or cross-section of the rocks. The layering in the rocks of the Assynt district is generally tilted gently toward the southeast, and so the natural line to choose for a cross-section would run roughly in a southeasterly direction. In a cutaway view of the rocks along this line, the different rocks appear as a vertical stack of gently tilted layers. If we think of the rocks in terms of a color code, then we can code a basic sequence of rocks pink, grey, green, and blue in order of decreasing age. These were the colors Peach and Horne used on their map: pink for the metamorphic rocks in the western region, grey for the white sandstones, green for the shales, and blue for the limestones. The eastern metamorphic rocks were color coded yellow.

A Giant Break in the Crust

So what does the color-coded cross-section through Peach and Horne's map look like? To begin with, the basic sequence of rocks, from pink to blue, is repeated twice, with the yellow (metamorphic) rocks right on top. This made no real sense to a geologist who was thinking only in terms of the steady accumulation of sediment in layers. It seemed impossible that the simple accumulation of sedimentary layers could lead to such a perfect repetition, even to the extent of containing the same fossils. But what at first is far more puzzling is the position, in the middle and at the top, of the pink and yellow metamorphic rocks. Explaining the presence of pink metamorphic rocks at the bottom of the sequence, in the west, was easy—they had been cooked-up and exposed at the surface before the sedimentary rocks were deposited. But, how could metamorphic rocks also be above the grey, green, and blue sedimentary layers?

The explanation that Peach and Horne came up with required large-scale horizontal movements of the Earth's surface. They proposed that the sequence of rocks could be explained if the deeply buried yellow metamorphic rocks had been pushed up and over both the shallow-water sedimentary rocks and their underlying basement along a gigantic, gently inclined break or fault in the Earth's crust. In the process, the rocks had been broken and fractured along numerous smaller faults, jostled and shuffled like a pack of cards so that the layers were placed out of order. In this way, older bits of the sedimentary sequence and its metamorphic basement ended up on top of younger bits. The extraordinary thing about the movement along the main fault is its sheer scale. To explain the repetition of the layers and the presence of metamorphic rocks right at the top, the horizontal movements had to be at least several tens of kilometers, and possibly over a hundred kilometers. Geologists call such a fault a thrust.

Peach and Horne's boss at the Geological Survey, Sir Archibald Geikie, immediately saw the importance of this thrust, "the discovery of which has made the north-west of Scotland a classic region for the study of some of the more stupendous kinds of movement by which the crust of the Earth has been affected." These stupendous movements became even clearer when Henry Cadell, a fellow geologist at the Survey, carried out some experiments in 1888. Cadell compressed horizontal layers of clay in a gigantic screw vice—there is a photograph in the Survey archives of him standing by this device, mustachioed and gripping the levers of the screw in a rather theatrical pose. He found that as he closed up the jaws of the vice, the layers or beds naturally piled up in a very similar manner to the rocks in Scotland, with "a major thrust or 'sole,' inclined at a low angle, along which the whole mass has travelled for considerable distances."

Cadell's experiments had shown not only that Peach and Horne's fault was to be expected when rock layers were pushed together or squeezed horizontally, but also that these movements, because they piled or heaped up the layers, were a way to thicken up the crust. Here was one mechanism to make the sort of deep roots of

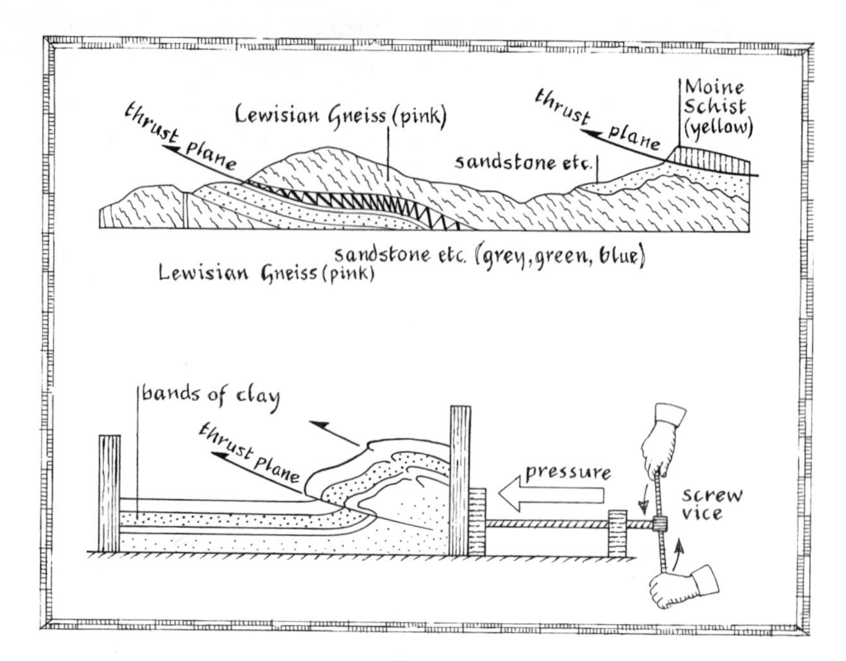

Stupendous horizontal movements in the Earth's crust: geological mapping at the end of the nineteenth century, in the Assynt district of northwestern Scotland, showed that the rock layers had been thrust up and repeated along gigantic faults. Early experiments with layers of clay in a screw vice managed to produce very similar structures.

mountains that George Airy had proposed in the 1850s when he had tried to solve the problem of the plumb line in India. It is clear that this was not lost on the geologists at the Survey. Cadell had "no doubt that, when other mountain systems come to be examined in the light of the researches by my colleagues and myself in the NW Highlands, and in the laboratory, these structures will prove to be of common occurrence." Prophetic words. But the mountains in Scotland must have been worn away long ago, because the hills here, as Peach and Horne recognized, have nothing to do with movement on their fault, but were sculpted out of the landscape by the waxing and waning of giant ice sheets much later on.

In the Doldrums

We have arrived at the stage in our story when the ship of geological ideas about mountain making is on a fair course, with a strong wind behind, plowing its way through the water to a new scientific world. The early stages of this voyage had been slow as the crew left their home port and made their way through a maze of confusing coastlines to the open ocean. But now, with sea all around, there seemed nothing to stop the ship reaching its inevitable scientific destination. The work of the early surveyors and geologists could be brought together into a unifying picture: George Airy's mountain ranges with their deep roots were the result of gigantic horizontal movements, heaping up or thickening the crust of the continents; the history of the Earth was certainly long enough for a whole succession of mountain ranges to have been raised this way, and these upheavals, as Charles Darwin had experienced during the earthquake in Chile, were still going on today.

The logical conclusion of this mobility—evidence for which had now been found, not just in Scotland, but in the European Alps, and much of Central Asia—was the idea of wandering or drifting continents, proposed by the German meteorologist Alfred Wegener. This theory was graphically summarized in 1920 at a meeting of the Royal Geographical Society in London by one of Wegener's supporters: "The surface of the Earth is like an ice-floe composed of solid blocks which are always crumbling against one another. These blocks as a whole seem almost immune from any change, but in between there are very energetic movements."

In Wegener's view, mountains were the result of that energetic movement when continental blocks collided. It was not a huge step from this to the theory of an Earth covered in a mosaic of restless tectonic plates—plate tectonics—the paradigm of modern geology. But then something strange happened to the ship of geological ideas. It became becalmed in the doldrums, and the crew started to argue among themselves. Nobody could think of a reason why the continents should be wandering, let alone colliding, and geologists began to lose confidence in their basic observations. Surveyors, fixing the great landmass of North America in the early part of the twentieth century, decided to follow Airy's rival, John Pratt, viewing the great ranges of the American Rockies as regions where the crust was less dense than normal. Recalling our previous analogy for Pratt's theory, they thought of the Rockies as well-risen bread compared to the heavy dough of the crust beneath the great plains farther east. It was hard to see how squeezing the crust beneath the mountains could result in this sort of variation; if anything, it required an expansion of the rocks. There were plenty of other confusing observations.

Geological investigations in many great mountain ranges on Earth, such as the American Rockies, Alps, Himalayas, and Andes, had begun to reveal large bodies of granite and other volcanic rocks. Perhaps the effects of the folding and faulting of rocks had been overestimated or misunderstood, and perhaps the creation of mountains is more directly linked to the forceful rise of buoyant and once molten granite as part of some deep form of volcanic activity, pushing the rock layers to one side? In this case, the contortion in the rocks would be a consequence, rather than a primary cause, of their uplift. In the 1920s the British geophysicist Harold Jeffreys, who had written what was considered to be the definitive work on the nature of the Earth's interior, distilling the study of literally thousands of earthquakes, was adamant that the Earth was far too strong—literally as rigid as steel—to allow large-scale horizontal movements in the crust. This seemed to rule out the possibility of drifting or colliding continents. Worn down by these theoretical arguments and conflicting ideas, geologists began to back down on advocating such an extreme mobility of rocks. Even if these movements had occurred, it was only in the remote geological past, at an early stage in the planet's history, when the Earth was very different. Darwin's earthquake was completely forgotten, and Charles Lyell's maxim that the present is the key to the past, so influential in the nineteenth century, no longer seemed profound or attractive.

This is probably a good place to leave the people on board that becalmed ship to bicker as they drift around in ever decreasing circles—let us hitch a ride on a passing albatross. The ship will be in these doldrums for many decades, not only rudderless but enveloped in thick fog—a strong puff of wind will be needed to get it back on course. However, we have enough ideas to start making sense of my first visit to Bolivia. The fact that this was 106 years after Peach and Horne had begun working on their map of the northwestern Highlands, 154 years since Darwin's experiences in Chile, and 254 years after the French surveyors made their measurements in Ecuador is a good illustration of how much time and human effort has gone into trying to understand our planet. Despite this, there was still geological confusion, clinging like wisps of mist to many of the high Andean peaks.

⁜ CHAPTER THREE ⁜

A Geological Reconnaissance

Reconnaissance, ***n.*** Examination of region by detachment to locate enemy or ascertain strategic features; preliminary survey. (*OED*)

MEMORIES OF MY FIRST visit to Bolivia are rather like a collage of pictures—some images bright and central, sharply in focus, others dim and vague, lurking at the edges of my mind. The purpose of this visit was to gain an overall impression of the Andes—at this stage I did not intend to carry out any specific geological research. And as I tried to record my impressions, I came to appreciate more and more the skill with which the young Charles Darwin had made sense of a vast panorama of observations when he began to explore the Andes on the voyage of the *Beagle*. Darwin recorded his observations and impressions in his journal. This journal makes fascinating reading today, with its jumble of encounters with local people, logistical problems, and off-the-cuff ideas about his observations. My trip, of course, was much shorter, and far less of an exploration into the unknown. I still have the notebooks I filled on this trip, with their mud-stained pages full of sketches and geological notes. When I leaf through them today, forgotten parts of the journey are suddenly brought back to life. And, as it turned out, the trip proved crucial to the success of our project, laying the foundations of almost all of our subsequent work. But it was not until many years later that I fully understood the significance of many of my observations.

I left Oxford for the Bolivian Andes in the Northern Hemisphere spring of 1989. I had a checklist of geological observations I wanted to make, strongly influenced by the conflicting ideas about their origin put forward in the scientific literature. I knew that there were several other geological teams or projects working in the Andes. These tended to divide up along national lines, staking research rights to large tracts of mountainous territory—later, I was to become aware of border disputes in which it was perceived that one group was trespassing on another's property. There was no end of American geologists and geophysicists from many universities—not far, I suppose, from their own backyards—with the biggest project based at Cornell University in New York and focusing on northern Argentina. There were French geologists in Peru, Germans and a few English in northern Chile, and, of course, the universities and geological surveys of the South American countries themselves. But the vast spaces of Bolivia seemed virtually empty of geologists. I hoped that our Oxford project would find a niche for itself here.

I was on the lookout for any geological evidence of horizontal and vertical earth movements. And I was particularly interested in when these movements had occurred. Evidence for such movements would lie in both the rocks and the landscape. I was also interested in signs of deep volcanic activity that might be related to mountain building—for example, the presence of large granite bodies that had been intruded into the crust deep below the surface volcanoes. Finally, John Dewey had asked me to find a field project for a new doctoral student whom we hoped would join our project.

Arriving in Bolivia

The air crew announced that we were about to land at the city of La Paz, Bolivia. We had swooped over the high spine of the Cordillera Real, with its snow-capped peaks, and were flying along the wide, barren, brown expanse of the Altiplano—literally, the high plain—of the Bolivian Andes. The ground was a mosaic of fields, low stone walls, and small piles of stones, heaped up to clear the land. The airport of La Paz, which is called El Alto (the high place), is perched on the edge of the Altiplano at 4,000 meters, making it the highest international airport in the world. The city of La Paz lies in an erosional bowl, just to the east, where rivers, flowing off the Andes, have nibbled at the edge of the high plateau, cutting deep gulleys. The city came into view, over the edge of the Altiplano, as the plane circled in preparation for its landing approach. I looked out to see a cluster of high-rise buildings surrounded by a vast expanse of low housing creeping up the steep sides of La Paz. The city spreads out both upward and downward—unlike most big cities, where the affluent suburbs are high up on hills, in La Paz they are at the lowest altitudes, over 600 meters below the level of the airport, where the air is less thin.

My ears popped and I gasped for breath as the oxygen in the aircraft seemed to disappear—the pilot was equalizing the cabin pressure with that outside. On a subsequent flight into La Paz I monitored the pressure difference with an altimeter. In the space of about five minutes, as we came in to land, the needle of the altimeter whirled round from a reading of about 2,000 meters (a typical equivalent altitude to the in-flight cabin pressure) to 4,000 meters—a rather brutal way of going to high altitude. At the altitude of El Alto, planes must land twice as fast as normal on a runway that seems to go on forever. In the 1970s a Pan Am jet had failed to land, missing the runway and flying straight into the side of a mountain. Since then, only specially trained pilots are allowed to land at El Alto.

As I emerged from the aircraft I was chilled by the thin, subzero air but dazzled by the sun reflected off the high, snow-covered peaks of the Cordillera Real, extending north from La Paz along the eastern margin of the Altiplano. I had to remind myself to breathe deeply, taking in lungfulls of air as we filed across the tarmac toward the arrival hall. All my fears during the flight of coping with the Andes flooded back as I desperately tried to rehearse the few words of Spanish I had prepared. With the help of a Spanish friend in Oxford I had concocted two "really useful" sentences: "¿Hay alguna persona que habla ingles?"—Does anybody here speak English? And "Mi trabajo es ciencia pura"—My work is pure science. The latter sentence was designed to damp out any suspicion that I was a spy, while the former sentence was intended to be a standard opener when dealing with official organizations.

Once through immigration, where I was challenged by a stocky Bolivian official in rapid Spanish, but who in the end seemed to be satisfied with my passport, stamping it "Entrada El Alto, 90 Dias" with a resounding thump, I emerged in the airport luggage arrival hall.

The journey from the airport to the center of La Paz is a stunning experience. The road winds down the sides of the steep edge of the Altiplano, descending over 400 meters. Below you, the city of La Paz is spread out in a bowl with the backdrop of the high Cordillera Real. In the early evening, this view is particularly dramatic, with the dark masses of the mountains towering above a myriad twinkling city lights. As the taxi swoops ever lower on the windy road, almost like a bird, the street life of the city begins to engulf you.

I ended up at the Hotel Gloria, in the old center of La Paz. After inspecting a number of dingy and poky rooms, I had finally managed to get one with a "vista" of Illimani—the 6,500-m ice-covered peak at the southern end of the Cordillera Real, which dominates the city of La Paz. With a wet handkerchief over my face, in an attempt to create some humidity in the air—the air was so thin and dry that it was painful to breathe in—I lay down on my bed and drifted off into a light and fitful sleep, wondering how I was going to have the energy to do anything now that I was here.

The next morning, as always, my prospects looked brighter. The first task was to make contact with a Jesuit priest called Padre Ramon Cabre, whose name I had been given by a colleague in Oxford. Ramon ran the last Jesuit mission to South America—a small remnant of the once vast missions in the lowlands of eastern Bolivia during the eighteenth century that threatened to form a state to rival the Portuguese possessions in Brazil. His mission was the Observatorio San Calixto, set up in 1913 to monitor earthquakes in Bolivia. He had contacts with many American universities and organizations, and the priceless gift, from my point of view, of being able to speak good English. I felt that he would be sympathetic to the aims of the research project and would be able to help me get the appropriate permissions from the relevant Bolivian organizations—and, indeed, all these things turned out to be the case.

Out of breath and breathing heavily, I met Ramon Cabre in his small office at the Observatorio, having struggled up hill from the Hotel Gloria. The Observatorio occupied an old palace that, like many of the original Spanish colonial buildings in La Paz, was situated on the high side of town and built like a fortress. From the street, all one could see was a high façade with small, barred windows on the first and second floors, and an immensely strong door in an archway at ground level. Yet, to me, the street aspect seemed strangely familiar, rather like that of one of the small medieval Oxford colleges. Once inside, the hustle and bustle of La Paz seemed to fade away as one entered the private world of a cloister surrounding a courtyard with a marble fountain.

Ramon greeted me, wearing the long black cassock of a Catholic priest. He immediately suggested that I should visit the Bolivian Geological Survey, known by the acronym GEOBOL (Bolivians love acronyms or abbreviations), and offered to take me there himself. We descended the steep streets of La Paz, Ramon striding ahead as we side-stepped the street sellers with their wares—usually sweets, penknives, and dubious-looking gadgets in soiled boxes—laid out on blankets across the pavement. There was a strong smell of urine in the air, mingled with diesel fumes from poorly maintained truck and bus engines. The Bolivian Geological Survey occupied a warren of a building in a narrow back street of old La Paz. The director was away, somewhere overseas. I would come to learn that this is where directors of large Bolivian organizations often seem to spend their time. However, the number two in the Survey, who was always referred to as Freddie, would be happy to see me. Much to my relief, Freddie had worked in the United States and spoke a little English himself. He welcomed the prospect of collaboration with the Universidad de Oxford—a very famous university, he assured me, and well known in Bolivia.

He immediately asked me if I was interested in an office—and perhaps a secretary too? Also, if I needed to travel anywhere, they had vehicles I could use. All this, of course, for the minimum amount of money—merely enough to cover their basic costs. He graphically illustrated the tiny amounts of money involved by defining a small gap between his thumb and forefinger. And if there was anything else I wanted, I should let him know immediately. I was rather taken aback by these offers. Usually, official organizations seem to pride themselves on the sheer impossibility of extending their facilities to outsiders, or even to their own personnel. I subsequently found out that the Survey had virtually no money whatsoever. The Bolivian government, on the advice, it was said, of the American ambassador (who many thought was the real president of Bolivia), had drastically reduced the annual inflation rate from 24,000 percent down to a respectable 8 percent by the simple expedient of cutting all government expenditure. This meant that most government organizations were in a state of limbo. They existed with a nominal staff on their payroll, but nobody actually got paid or did anything. Therefore, in these circumstances, I suspect that Freddie could not believe his luck when he saw me come through the door. Perhaps he saw me primarily, not as a geologist from Oxford University, but a walking bundle of dollar bills?

It became clear in my conversation with Freddie that the geologist who could help me most with advice on the geology of Bolivia was Raul Carrasco. Unfortunately, Raul was away at the moment doing some private consulting work—this was the only source of revenue a Survey geologist could expect to have in the present dire financial situation. However, it was rumored that Raul would be back next week. Perhaps I could come back then?

Traveling with Raul Carrasco

Raul Carrasco was a tall and quiet-spoken Bolivian geologist who had once been director of the Geological Survey. He had worked together with a number of foreign geologists on various map-making projects and probably had the best general knowledge of anybody on the geology of Bolivia. When I first met him he had just got back from advising a construction company about the stability of part of the main road from La Paz to Cochabamba. Cochabamba is the second largest town in Bolivia, situated at an altitude of

2,500 meters in a deep valley in the heart of the rugged Eastern Cordillera, about 200 kilometers, as the crow flies, southeast of La Paz.

Raul immediately suggested that he accompany me on a trip through the northern Altiplano, and then on to Cochabamba, crossing part of the Eastern Cordillera. He thought this would provide a good introduction to the range of rocks in the Bolivian Andes and might help me decide where to focus our research project. He also strongly urged me to draw up a formal document with the Survey, setting out how we proposed to work. This document, called a *convenio*, is a prerequisite of any sort of project in Bolivia. It commits nobody to anything but provides the basis for all subsequent official permissions. With some advice from Ramon Cabre, I drew up a convenio in my hotel, written on large white hotel paper napkins (these were the only large sheets of paper I could lay my hands on), while Raul started making arrangements for our projected trip.

When I met up with Raul the next day, I discovered there was a small problem. The survey did not have a vehicle that was in a fit condition to travel out of La Paz. Would I be prepared to pay to have the car fixed? I was a little surprised, but I guessed that this would be part of the cost of using their vehicle. However, there was another small problem. Could I give the driver some money so that he could buy some petrol to put in the vehicle so that they could drive it to the mechanic? And could I pay the mechanic's bill myself? I realized for the first time how dire the financial circumstances of the Survey really were. However, with a little financial lubrication, the car plus driver and Raul arrived at my hotel early one morning. The driver was called Manuel, and the car was a blue, short wheel-base Toyota Land Cruiser, which could only just fit three people and their luggage. On the roof was tied a large white plastic container. I thought at first that this was for petrol, but it turned out that Manuel was hoping to buy a large quantity of *chicha*—a local brew made from maize and (it is proverbially rumored) urine—in Cochabamba, the chicha capital of Bolivia. The state of the spare tire, bolted to the rear door of the jeep, was a particularly ominous sign—it was so worn that the inner tube was ballooning out of holes in the tread.

Down the Zongo Valley

The first part of the journey was dedicated to visiting the curiously named Zongo Valley in the Cordillera Real, just north of La Paz. I still feel that a name like this really belongs to some lost African tribe or kingdom. The Zongo Valley is renowned for the remarkable change in elevation along its length. The British government built a major hydroelectric scheme here in the 1960s, exploiting this change in elevation. Numerous aqueducts, tunneled into the mountainsides, channel water into electricity generating stations at various points down the valley. The lowest generating station is nearly three thousand meters below the top one. From a geological point of view, the valley has another interest. It cuts deeply through huge bodies of once molten granite.

We took the road north of La Paz, which winds up into the Cordillera Real, above the snow line and through a pass at an altitude of 5,200 meters. The ice-covered mountain of Huayna Potosi towers above the road to an elevation of over 6,000 meters. From here, the road, cut into a granite cliff face, plunges into a cirque carved out by a mountain glacier during the peak of the ice age. And from then on it is down, down, down along a rutted and bouldery track. This road provides an extraordinary chance to witness all the changes in landscape, flora, and climate down the sides of a mountain. The top of the Zongo road is in a landscape of ice-smoothed rock faces, ice fields, and bouldery glacial moraines. Here, the air is so thin there is only half the amount of oxygen at sea level. The bottom of the Zongo road, which we reached a few hours later, is in dense tropical jungle at about the same elevation as the Scottish highlands. The air is thick and humid, with dense plantations of banana and other fruit trees covering the hill sides. The contrast cannot be greater. It is hard to believe that you are still in the same country.

On the journey down I had taken a keen interest in the outcrops of granite, getting Manuel to stop the car frequently while I squeezed myself out to have a look. It was good to wield my geological hammer again, searching for angular corners of an outcrop and then giving them a smart blow. I felt for the first time that I had truly embarked on field work in the high Andes. If you have not tried looking at rocks in the field yourself, a geological hammer may seem a rather odd piece of equipment. In fact, geological field work without a geological hammer is a bit like bird watching without a pair of binoculars. Rocks are difficult to see. They are weathered, stained with mud, covered with lichen, and awkwardly shaped. But with one blow of a geological hammer, the rock reveals its fresh face, glinting with cleaved crystal surfaces. And there was no shortage of granite in the Zongo Valley to hit like this. I collected some samples in the hope of being able to date, back in England, when the rock had solidified from a molten state. Perhaps these intrusions had played a role in building up the present mountains? (It turned out that they had not because they were far too old.)

At the lowest point of the Zongo Valley road we left the jeep and scrambled along the river, which had followed us down. By now it was a raging torrent, half blocked by enormous boulders, the size of houses. The water-smoothed bedrock was a grey-blue slate, spangled with beautiful white cross-shaped crystals, each the size of a small piece of jewelry. This mineral is a form of aluminium silicate, commonly called chiastolite, and typically grows in this type of rock when the temperature exceeds a few hundred degrees centigrade. Between the chiastolite crosses, small golden cubes of pyrite caught the sunlight, giving the rock a speckled appearance. As the cool mountain waters rushed past us with a roar, and monkeys cavorted in the forest, we were peering into the bowels of the Andes.

The journey down the Zongo Valley had turned out to be a journey deep into the crust beneath the mountains. The crystals in the rocks at the end of the road could only have crystallized at the temperatures and pressures that exist over ten kilometers beneath the Earth's surface. The fact that these crystals were now at the surface is testimony to the power of the Zongo River and its many predecessors to wear away the landscape, carting vast quantities of rock away to the lowlands. The relentless rise of the Andes must have continually counteracted these forces of erosion, ensuring that the Zongo Valley, with its staggering drop in elevation, exists today.

Visiting the Altiplano

We made the slow and tedious journey back up to the Altiplano, grinding along the rough road in low gear. Gradually the forest thinned and gave way to alpine scrub above the tree line, now shrouded in mist. I began to develop a sharp headache as the air thinned. Eventually we emerged over the lip of the Cordillera Real and saw the brown, barren plains of the Altiplano stretch into the distance before us. On the horizon, silhouetted by the setting sun, high, conical volcanic peaks stood like sentinels along the border with Chile.

The Altiplano is an extraordinary feature of the Earth's surface. It is a vast high plateau, bigger than Britain and at an average altitude of about 4,000 meters. It is hemmed in by the giant volcanoes in the west—the Western Cordillera—and the peaks of the Eastern Cordillera, with its high spine of the Cordillera Real, in the east. Large portions of it are virtually flat and featureless, punctuated only by small, conical hills or ephemeral lakes and salt pans. The landscape belies the altitude—it is a treeless desert, covered in a low, scrubby vegetation, which the locals call *tholla*, with resinous branches that make excellent firewood. Wide, sandy rivers or dried up stream beds traverse the plains, snaking their way toward the shallow lakes or salt pans that occupy the lowest points. The rugged peaks and valleys of the Andes seem far away, yet, on the Altiplano, you are higher than the average elevation of these peaks and valleys. Between the months of June and August, the Altiplano can be incredibly cold at night—temperatures often reach −20°C—but unbearably hot in the day. I have camped out in the Altiplano during these months to wake up with a white hoar frost coating my sleeping bag and the inside of my tent, my water bottle frozen solid. The only similar environment on Earth is the high plateau of Tibet.

We made for the village of Patacamaya, which lies at the junction with the main road to Chile. Here, in a small, mud brick room of a *residencial*—Bolivian-style traveler's hostel—I spent the night, huddled in my down sleeping bag in the subzero conditions. The hostel had no running water, and soon after sunset the only drinking and washing water, stored in old tin oil drums, began to freeze. The next morning I needed my geological hammer to break the ice. We set off after an Altiplano breakfast of *api* and *pastellas*—a hot drink, a bit like liquid porridge, served with fried batter puffs. I liked the puffs, but not the api—there was something uncomfortable about its consistency. We turned off the main road into the dusty Altiplano near the Chilean border, heading for the mining town of Corocoro. Corocoro in its heyday had one of the largest deposits of native copper in the world. The copper formed sheets of metal within the rock, requiring virtually no processing to be usable.

Our primary reason for visiting Corocoro was not to see the copper but to examine the thick sequence of sedimentary rocks that hosts it. These sedimentary rocks are sandstones with a distinctive red color—for this reason, geologists call them red-beds. The red color is due to minute grains of iron oxide—literally rust—within the rock. The sediments most likely to rust in this way are those deposited by rivers and exposed to the oxygen-rich atmosphere. In fact, mining geologists had estimated that a pile of red-beds nearly eight kilometers thick had been laid down in the region around Corocoro. Such an enormously thick sequence—I know of no other sequence like this in the world—implied a very large system of rivers carrying detritus from a mountain range to the lowlands in what is today the Altiplano. But where were these ancient highlands, and why were the lowlands now at an elevation of 4,000 meters? Here was a geological puzzle to get my teeth into. What was particularly intriguing was the evidence that the Corocoro red-bed sequences were, geologically speaking, fairly young. They certainly were younger than the Cretaceous period, when dinosaurs dominated the Earth—the red-beds lay on top of Cretaceous rocks containing dinosaur footprints. Thus, understanding the origin of the Corocoro red-beds could provide vital clues to the origin of the Andes themselves.

Corocoro also sticks in my mind as the scene of my first Bolivian tire puncture. Manuel stopped the car with much muttering and got out to look at the damage. The tread of the back wheel had been pierced by a long rusty nail. Manuel changed into a pair of overalls and squatted down on the dusty road, jacking up the car while Raul and I examined a roadside outcrop of rock that contained some unusual pebbles. As usual, the wheel nuts had been put on much too tightly and required an incredible leverage with a spanner before they would loosen. Raul did not seem inclined to help while Manuel strained away, so I decided not to interfere. Eventually the punctured tire was replaced with our spare tire, which was almost in a worse condition. Despite extensive inquiries there seemed no hope of having the tire patched in Corocoro. We were faced with the prospect of getting back to Patacamaya along rutted tracks with our ballooning spare tire. We made the mistake of trying a shortcut, or *atajo* as Bolivians call them. My subsequent experience in Bolivia is that all shortcuts are long cuts, and this proved to be the case as our track seemed to melt into the trackless wastes of tholla scrub in the Altiplano.

Manuel frequently tried to get directions from startled men and women who were keeping an eye on their flocks of sheep as they sheltered in the shade of a bush during the hot afternoon. They usually replied in the local Quechua language, indicating the way with delicate fluttering motions of the hands. Some years later, in similar circumstances, I remember the driver shouting to a young Bolivian woman for directions to some village in the Altiplano. She immediately started running away from us. The driver got out and ran after her, speaking in a mixture of Quechua and Spanish, calling out to her: "Doñita, doñita (little lady, little lady), please can I speak to you." Eventually, she turned round and shouted back in Quechua, saying, so the driver told me later, that she was far too busy to stand around talking and had many more important things to do. At which point she ran off.

To everybody's relief, we finally emerged on a narrow track that looked at least like a definite road. Manuel exclaimed with joy: "Este es el camino principal! (It's the main road.)" Back in Patacamaya, I insisted that we buy a new spare wheel. After another freezing night in a rather "iffy" residencial—the sort of place where you don't really want to touch anything—we followed the real main road south to where it turns eastward across the rugged Eastern Cordillera to Cochabamba.

The Eastern Cordillera and Cochabamba

The road across the Eastern Cordillera to Cochabamba is a remarkable feat of engineering. It runs in a roughly east-west direction, traversing a series of northwest trending ridges and valleys. The highest point on the road is about 4,500 meters above sea level, and the lowest point, Cochabamba, is at an altitude of about 2,500 meters. The central section was still a narrow and rough dirt road, but the final descent into Cochabamba had been recently widened and tar-sealed. This involved huge, steplike road cuttings on the steep mountainsides. Sections of the new road had already broken up and collapsed where they had been cut into ancient but still unstable landslides. But the new road clearly revealed the bedrock; the edges of the dirt sections were caked with a thick layer of fine dust, stirred up by the traffic of heavy trucks.

The Eastern Cordillera consists of range upon range of stark, rocky ridges separated by deep valleys; the whole landscape is dry

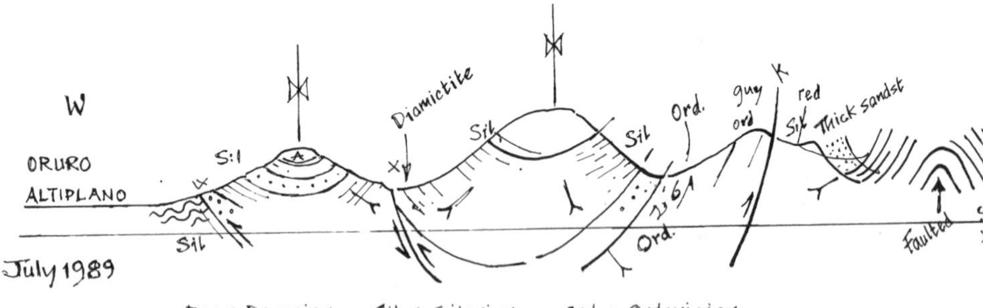

Rock layers in the landscape: a page from my field notebook. A rough geological sketch, made on a journey across part of the high Bolivian Andes,

and treeless. The grain of the country has been determined by the grain of the strata in the underlying bedrock. As far as the eye can see, the mountainsides are covered with a patchwork of fields cultivated by the native Quechua or Aymara peoples over the centuries. The only access is along mule tracks that zig-zag up and down the mountains. The fields are cunningly positioned in gulleys or on small ledges, designed to catch as much as possible of any rain water flowing off the mountains. Only in this way, with a skill that must have been acquired over many generations, do the farmers manage to nurture crops of maize, wheat, and potatoes in this high and desolate region.

As we made our slow way along the winding road, I attempted to sketch the geology in the road cuttings, producing a more-or-less continuous log of both the types of rock and the attitude of the layers. The rocks consisted mainly of thick, monotonous sequences of sandstone and shale. Raul told me that fossils collected in these rocks—buglike creatures called trilobites—showed that they had been laid down in the Ordovician and Silurian periods of geological time, over four hundred million years ago. It was strange to be in South America and use divisions of geological time based on rocks in Wales and named after ancient British tribes. The once horizontal rock layers were now tilted at steep angles. By carefully recording the direction in which they were inclined, sometimes to the northeast, sometimes to the southwest, it was possible to discover great folds in the layers of the bedrock. Occasionally, distinctive bands of rock were broken and shifted across faults in the mountainside.

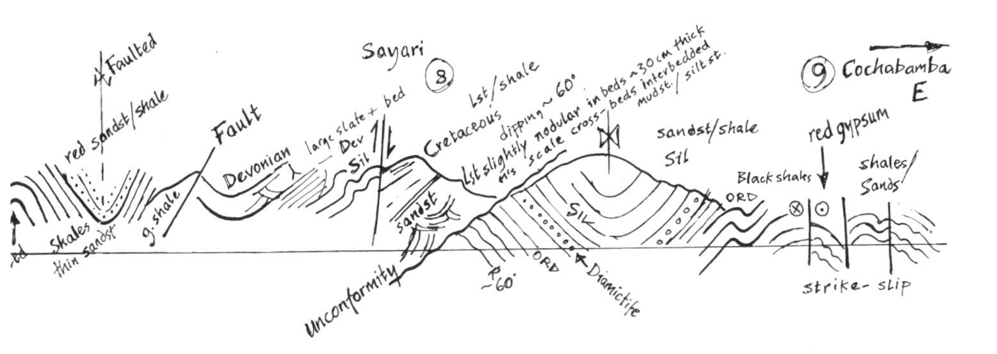

reveals great folds and faults in the rock layers—clear evidence of horizontal squeezing of the crust beneath these mountains.

The folding and faulting were clear evidence to me of horizontal squeezing and shortening in the region. Was this the event that had pushed up the Andes? The rocks in the road cuttings suggested an answer.

Just before we started the last big descent toward Cochabamba, we crossed a ridge called Sayari. Here, a prominent band of white limestone runs along the ridge crest. Raul said that these were deposited in the Cretaceous period and were typical of the rocks of this period throughout the Bolivian Andes. These limestones, with their interleaved layers of shale, were deposited in a huge lake or inland sea that existed sixty-five million years ago, right at the end of the Cretaceous and just when the dinosaurs went extinct. These rocks were therefore proof that the dinosaurs never saw the Andes because these mountains had not yet come into existence. However, the once horizontal Cretaceous strata are now steeply tilted, clear evidence for subsequent large-scale movements of the crust that may have played in role in the rise of the present mountains.

We could see the valley of Cochabamba far below us for a long time before we actually entered it. The valley is a curious feature because it forms a wide, flat plain deep within the Andes, running in an east-west direction across the general grain of the main mountain ranges. The northern side of the valley is dominated by the high peaks of the Tunari mountains (Cordillera Tunari). In fact, the Cochabamba depression has the characteristic features of a gigantic rift or split in the crust that cuts right across the mountains. In this case, the crust has been pulled apart, not squeezed together—the opposite to the movements in the crust needed to build the mountains. It struck me that unraveling the geological story of this valley would make a good project for a future thesis student. And so was born Lorcan Kennan's research project, which he began working on the following year.

The old town of Cochabamba consists of white single- or double-story Roman-villa-style houses with internal courtyards. The streets are arranged in a grid pattern with the main plaza (*plaza principal*) at the center—here much of the commercial life of the town thrives. I was told that the hyperinflation in Bolivia during the early 1980s was actually triggered by the money changers in Cochabamba. It seems that a rich Bolivian family was selling a valuable property in Cochabamba at a time when the Bolivian government had decided to make commercial transactions in U.S. dollars illegal. The property, worth several million dollars, had therefore to be paid for in local pesos. Immediately, the sellers tried to convert the pesos into dollars on the black market, using up virtually the entire local supply of dollars. Now that there was effectively no U.S. dollar backup to the local currency, confidence in the peso collapsed, triggering hyperinflation as people frantically tried to convert their money into what few dollars were available. At the height of the hyperinflation, money changers would calculate the exchange rate by the hour. Raul told me that the first thing he did at this time, when he got paid, was spend all his money, and he often had to resort to a system of barter. I felt that his quiet and modest way of speaking played down the real suffering that he and many other Bolivians were still experiencing.

The next day we crossed the wide, flat valley bottom and climbed the rough road into the Cordillera Tunari. I was looking for evidence of movement along fault lines that may have created the Cochabamba basin. The northern margin of the valley is extremely steep, running in a gentle arc. Raul thought that this might be a fault line where a steep fracture in the crust reaches the surface. Movement during earthquakes on such a fault may have progressively dropped the base of the valley. Rivers running off the Cordillera Tunari had deposited gravel in huge, conical fans on the valley margin, slowly filling up the rift. Further evidence for this faulting might lie in the hot springs of Urmiri, perched on the steep valley side and possibly bubbling up from a depth of several kilometers along the rift fault. Before the Second World War these hot springs had been developed into a resort with baths and accommodation. The resort is now a ruin, with crumbling walls and rubble-filled baths. Only a large basin outside still has hot water flowing through it. The locals use the hot water for washing themselves and their clothes. When we visited the springs, the surrounding ground was gaily laid out with colorful striped blankets, left to dry in the sun.

One final place we visited in the Cochabamba valley was the village of Sipe-Sipe. Sipe-Sipe had been badly damaged in an earthquake in 1909. One church, in particular, had been almost completely destroyed—the bell tower is all that remains, standing forlornly at the edge of town. The village sits on a fault line that Raul thought may have ruptured during the earthquake. I could find no clear evidence of recent breaks in the ground, but this did not mean much because such breaks are often obliterated soon afterward, covering up the tracks of the earthquake. The frequent earthquakes in the region indicated that earthquake rents and active fault lines must exist somewhere. But where were they? I left Sipe-Sipe determined to find them.

Santa Cruz and the Lowlands

Back at the GEOBOL offices in La Paz, I discussed the proposed *convenio* for our project with Raul and Freddie. The general headings that I had hastily put together (written out on Hotel Gloria napkins) seemed to be acceptable. The project would investigate all aspects of the geology of the Bolivian Andes in all "sectors" (this would cater for every conceivable possibility)—the Altiplano, Eastern Cordillera, Western Cordillera, and eastern foothills, or Sub-Andes, as they are usually called. The latter two regions I had not yet visited. We agreed that Oxford University would pay for everything (this would ensure that the Survey had no financial liability), but there would be an opportunity for Survey geologists, at the Survey's expense, to join us in the field if they so wished. In addition, the Survey would act as our official sponsor, ensuring we had the correct permissions from both "civil and military authorities" to run our project in Bolivia. Freddie undertook to get the document drawn up in Spanish on official Survey letterhead. Then, it could be formally signed by the director of the Survey, when he got back from his overseas trip.

In the meantime, I decided to visit the eastern foothills of the Andes—the Sub-Andes—where the mountains first begin to rise up out of the Amazon rain forest. This was the most earthquake-prone part of Bolivia—always a good sign to a geologist—and was also the region of most intense and successful oil exploration. Freddie suggested that I hire a Land Rover from the local Survey office in Santa Cruz. Raul offered to set up a meeting with the Bolivian state oil company, known by the slightly intimidating name of Yacimientos Petroliferos Fiscales Bolivianos (Bolivian Oil Deposit Company), or YPFB for short. Santa Cruz to an inhabitant of La Paz has the flavor of Monte Carlo—a glamorous city full of exciting action. In 1989 most of the action revolved around cocaine dealing, worth more than a staggering billion U.S. dollars and nearly an order of magnitude more money than the total tax base of the official Bolivian economy. The most glamorous feature of Santa Cruz are the women, or so Freddie told me. Their lithe figures and blonde hair—thought to be the result of intermarriage between the Amazonian Indians and Spanish settlers—are in marked contrast to the stocky build and black hair of the high-country Aymara and Quechua women.

I flew down to Santa Cruz in the lowlands, or Santa Cruz de la Sierra (the holy cross of the mountains), as it is properly known. The flight lasts about an hour and took me right over the route in the Altiplano and Eastern Cordillera that we had so laboriously traveled the previous week. I could see the layers of rock tilted on end, following the ridges in the Eastern Cordillera, as well as the abrupt range front of the Cordillera Tunari along the edge of the Cochabamba valley. Gradually, as we flew eastward, the elevation of the valley bottoms became progressively lower, and a lush green vegetation began to cloak the mountainsides. About a hundred kilometers from Santa Cruz the ranges were enveloped in a dense bank of clouds—here the moist air from Amazonia condenses as the air flows upward over the foothills of the Andes. This region, known as the Chapare, is hot and humid with several meters of rain each year—an ideal climate for the coca plant, the raw material of cocaine.

As we landed I had a feeling of euphoria. I suddenly became aware of how stressful it had been at high altitude, living in an atmosphere with so little oxygen. The aircraft filled with the humid air of the lowlands, so thick that one could almost cut it with a knife. Never had air seemed so delicious as I breathed in this oxygen-rich diet. Also, the landscape looked wonderfully green after the brown Altiplano and Eastern Cordillera. I took a taxi into town, making for a middle-range hotel recommended by Raul. Like Cochabamba, Santa Cruz was experiencing a building boom, and the old colonial one-story houses with their internal courtyards and broad covered pavements had started to sprout huge, multistory glass blocks. Many of these were probably financed by the same people who lived in luxury villas with large manicured grounds on the edge of town—big players in the drug trade (*narcos*), the taxi driver assured me. As we sat in a traffic jam, listening to a cacophony of car horns and revving bus engines, I started to become aware of the unpleasant aspect of a hot and humid climate—I was dripping in sweat. I collapsed in my hotel room, grateful for the stream of cold air from the air conditioning unit.

I was woken early the next morning by the ring of the telephone. The voice on the line asked for Señor Lamb. That's me, I thought, rather surprised in my sleepy state to be receiving a telephone call in Bolivia. The caller obviously could not speak English, but I gathered from frequent mention of GEOBOL, *vehiculo*, and *la mañana* that somebody was trying to contact me about the GEOBOL Land Rover for this morning—I had discussed this with Raul and Freddie in La Paz, and obviously something was happening. And later that morning, an extremely battered, sand-colored Land Rover pulled up in front of the hotel.

The GEOBOL office in Santa Cruz had been set up by the British government in the 1970s as part of a project to assess the mineral resources of the lowlands of eastern Bolivia. This region was underlain by the oldest known rocks in Bolivia, dating from the Precambrian, a period in the Earth's history before multicellular organisms existed. The huge mineral deposits of gold, silver, lead, chrome, and nickel in Africa, Canada, and Australia occur in rocks of this age, and so it seemed reasonable to assume that similar deposits would be found in eastern Bolivia. Geologists from the British Geological

Survey had worked together with Bolivian geologists, spending several months each year in the jungle. Land Rovers, donated to the project by the British government, were a further symbol of British involvement. Later, when I was back in England, I contacted one of the British Geological Survey geologists who had been seconded to the project in the late 1970s. He told me that the field work was extremely difficult. Teams of geologists would set out with several Land Rovers, literally hacking their way through the jungle to get to outcrops, often using teams of men with chainsaws. The Land Rovers had to be pulled across deep or muddy rivers with winches anchored to a cable on the opposite bank. And the incessant rain and myriad insects made camping in the jungle particularly uncomfortable.

The result of this effort was a series of superb geological maps of eastern Bolivia, along the border with Brazil. When I visited the Santa Cruz offices, the British geologists had long since left, and the offices were virtually abandoned. It was quite a moving experience: the empty rooms still had notices and family pictures pinned up on the walls, exactly as they had been when the building was a hive of activity, full of busy geologists and secretaries. Chairs and desks were stacked up in the middle of the entrance foyer, and outside, the remaining Land Rovers were parked forlornly, without wheels, and in some cases, without engines. I have never seen such battered-looking vehicles—it seemed as though each vehicle had been attacked with a sledge hammer. In fact, this battering was an inevitable consequence of the appalling conditions in which the geologists and Land Rovers had to operate—a tribute, I suppose, to both British fortitude and engineering.

Before beginning my excursion into the Sub-Andes, I wanted to make contact with geologists in the Bolivian state oil company (YPFB). Raul had arranged for me to see Oscar, a Bolivian exploration geologist who had been working in the Sub-Andes for many years. Unlike the Survey, YPFB was a commercially successful organization. Their offices on the outskirts of Santa Cruz were buzzing with life. A receptionist called through to Oscar, who ushered me into his air-conditioned office. Oil companies tend to operate on a need-to-know basis. They also have access to a vast amount of information that they have acquired at great cost, and which they are very reluctant to divulge to anybody else. I always feel a bit like a child in a giant toy shop when I visit an oil company. The company geologists casually pull out maps and sections dripping with information interesting to a data-starved academic. However, if you show too much interest the maps are discreetly put away again, or you are referred to a senior manager who will, of course, be happy to let you look at them, but unfortunately he is away at the moment. And so it was with Oscar.

The Land Rover I had hired had been made operational only by cannibalizing the remaining vehicles in the GEOBOL compound. It sounded as though a large number of tin cans were being dragged along behind. This turned out to be due to a roll of wire that had somehow got wrapped round the back axle. I had a driver and companion geologist, neither of whom spoke any English. I resigned myself to communicating in sign language and pigeon Spanish as we headed west out of Santa Cruz toward the Andes. We followed the old road to Cochabamba, built by the U.S. Army in the early 1950s to provide a link between the highlands and lowlands of Bolivia. Prior to the construction of this road link, the lowland provinces were on the point of splitting away from Bolivia and becoming part of Brazil because they had so little contact with the rest of the country. Santa Cruz itself is situated on the flat plains, nestled against the edge of the Andes at an elevation of about 700 meters. Behind us, to the east of Santa Cruz, these plains stretched for hundreds of kilometers. Human activity since the city was founded in the late 1500s had cleared away much of the jungle, turning it into grassland suitable for ranchers; a pall of wood smoke still hung over the horizon. But in front of us, toward the west, a low, green hill rose abruptly from the flat ground—this is the first of range upon range of hills, then mountains, of the Andes.

Where the old Cochabamba road crossed the range front and climbed up into the hills, evidence for movements of the crust on a colossal scale was clear. The rock layers, steeply tilted and crushed and broken, were falling apart, bringing down the lush vegetation with them. It was obvious that we had crossed a major boundary in the rocks, a fault line that ran along the foot of the first ridge to rise above the plains. From then on the road climbed up through deep gorges with towering cliffs of rock strata, often dipping at crazy angles. I knew from the sections and maps that Oscar had shown me in the Santa Cruz YPFB offices that the rock layers beneath the plains were, like the plains themselves, virtually flat and horizontal. I was therefore deeply impressed by the clear link I was witnessing between mountains and the intense contortion of the rocks. Here was a clue, if ever there was one, to how the Andes had formed, or at least the mountains on this remote fringe of the great Andean ranges. I was more and more convinced that the hills had been pushed up by huge horizontal forces in the crust. I was pondering these ideas as the old Land Rover vibrated and rattled along the deeply rutted road. It was too noisy to talk, anyhow, and we were choked by a cloud of dust that had found its way inside, settling in our hair and on our baggage. Shaken, dirty, hot, sweaty, and tired, with the noise of the Land Rover ringing in our ears, we arrived at the village of Samaipata at an altitude of about 1,500 meters.

We spent the evening drinking beer and trying to communicate. The driver showed me pictures of his children. The geologist explained that he knew nothing about the geology of the mountains because all his work had been in the lowlands as part of the British mapping project. I did not expect to gain much on this trip except a feel for the foothills. In fact, just the experience of crossing the major fault line that separated the mountains from the plains made the excursion worthwhile.

The Pacific Coast and Atacama Desert

Time was running out on my first exploratory visit to Bolivia. I had one final job in Bolivia, and that was to get the GEOBOL convenio signed. At last, the director had returned and would be able to meet me. I was finally admitted into the director's office with its antechamber of secretarial staff. On crisp sheets of GEOBOL-headed paper, the terms of our convenio had been typed out. There were two copies—one for the Survey and one for me. After the initial introductions and explanations, the director went through the convenio, initialing each clause. At the end of the document we each signed our names with a great flourish. I was pleased to find that I had been promoted to professor (*catedrático*) of Oxford University. Then, finally, each page was signed and dated individually with the official stamp of the Survey. This precious document was our passport to geological research in Bolivia. In the years to come it would prove its worth time and time again with both civil and military authorities.

A visit to northern Chile, on the western side of the Andes, would complete my introduction to the region. Bolivian airlines operate a daily flight to the Chilean port of Antofagasta, which in the nineteenth century was once part of Bolivia. The large deposits of nitrates, which at one time were the main world supply of the basic ingredient of both fertilizers and explosives, provoked a challenge to Bolivia's claim. Far from their base in the Altiplano, and hopelessly outnumbered and ill-equipped, the Bolivian Army was defeated in the War of the Pacific in the 1870s when the Chilean Army occupied Antofagasta without warning. This had the catastrophic consequence that Bolivia lost its seaports and became a landlocked country, dependent on its neighbors for the import of all goods. As a consolation prize, the British government, which had interests in the nitrate deposits, built a railway from the coast to La Paz. However, Bolivia has never officially accepted this defeat, and there is a museum in La Paz dedicated to showing maps of Bolivia with the Pacific coast as part of its territory. "The sea is our right" is the motto of the military garrison in La Paz. So, from a Bolivian point of view, I was not really leaving Bolivia after all.

The landscape around Antofagasta is the most barren on Earth. Here, in the Atacama Desert, the land is naked rock—it rains once every fifty years. The desert extends from the coast to the high volcanic peaks of the Western Cordillera in the Andes, about fifty kilometers farther inland. However, there is a narrow range of hills along the coast, behind Antofagasta, which gives way to a wide inland valley at an altitude of about 1,000 meters, with the Andes themselves towering in the distance. A major fault line, known as the Atacama Fault, runs along the boundary between the valley and the coastal hills. Movement along this line has created a gigantic scar in the landscape running for hundreds of kilometers up the length of northern Chile; unlike fault lines in more temperate parts of the world, the dry climate of the Atacama perfectly preserves ground breaks during earthquakes. The Atacama Fault is well known to geologists throughout the world; it has long been speculated that some of the large earthquakes in this part of world are the result of movement on this fault. But, it is something of an enigma, because this movement seems to be raising the coastal hills and bringing the Andes, farther east, down.

The chance to examine the Atacama Fault was the main reason for my visit to Antofagasta. However, it was almost satisfying enough to just stand on the edge of the Pacific Ocean, with small waves breaking over my feet, and reflect on the fact that I had now traversed the second largest mountain range on Earth. The previous week I had been in the Amazon jungle looking westward toward the eastern foothills of the Andes. Looking out to sea, I was struck by the thought that only about seventy-five kilometers away there was a deep trench in the sea bottom, running like a gigantic groove along this side of South America. Here, a great slab of the Pacific Ocean floor was sliding beneath me, as it sank back into the Earth's interior. And not far behind me lay the great chain of active volcanoes. I was standing at the junction between two great tectonic plates, where vast forces within the Earth are at work.

Next morning I made my way to the nearest car rental company and took advantage of a special weekend offer to rent a car. Unlike Bolivia, where car rental was nonexistent and nobody ventured out of the towns in anything other than a truck or four-wheel-drive vehicle, suitable for the rough roads, Chile, in some respects, feels closer to Europe with a surface veneer of sophistication. Most of the main roads are tar sealed, there are cash machines and public telephones on street corners, and travel agents offer holidays in

Europe and the Mediterranean. All this lulled me into a false sense of security, which I was rudely woken from when I drove my small hatchback out of Antofagasta and into the Atacama Desert.

Tourists do not normally go to the places that geologists want to visit. Thus, the hatchback, designed for a tourist, was ideal for driving around Antofagasta or traveling on the main road north or south to other major towns. But I wanted to visit the Atacama Fault, and that involved leaving the frequented tar-sealed roads and venturing down rarely used dirt tracks. At first I did not really understand the full significance of this. However, as I left the main road and headed toward the fault, I began to wonder if what I was doing was sensible. I was completely alone in a strange country without any camping equipment or even much water. I drove on, trying to suppress these fears. After driving more than a hundred kilometers through the desert on a track that was in places so sandy that I either lost traction or hit the ground with the bottom of the car, I completely lost my nerve. When I stopped the car I found I was shaking and my hands were wet with sweat. I suddenly snapped. I was not going to do this. And so I gingerly turned the car around, terrified that I would get stuck in the sand off the track, and headed straight back to Antofagasta. In my urgency to get back to civilization, I found I was driving faster and faster, kicking up a long trail of dust. When I finally reached the tarmac main road, it seemed as though I had found a lifeline back to safety.

I had noticed another car rental company in Antofagasta that hired out four-wheel-drive vehicles. This was considerably more expensive than my weekend deal, but by now I did not care. The woman at the original rental company seemed surprised when I returned the vehicle so soon. "Why?" she asked. I explained that I was a geologist and the car was not suitable. She shrugged her shoulders and refunded me a small fraction of the money I had originally paid. I then went straight to the other company and hired their jeep. This time I was going to be more careful. I stocked up with several gallons of mineral water, some food, and camping equipment, and with a full tank of gas I headed back out into the desert.

My main concern was to determine the fundamental nature of the Atacama Fault. It had been thought to be either a rift-type fault, where the crust was being pulled apart, or a strike-slip fault, with the two sides sliding horizontally past each other. I wanted to see if any features of the landscape could be used to determine the most recent movements. The fault showed clearly as the sharp edge of a low ridge, a few meters high, that broke the surface of the desert. I found the ground firm enough that I could drive virtually anywhere, so I left the track and followed the ridge across country. I was searching for any stream where I might see the fault exposed in the banks. Eventually, after negotiating a number of tricky gulleys and ridges, I came across one; extraordinary as it might seem, the stream bed and banks had been sliced up during previous earthquakes, and the resulting disruption was perfectly preserved in the arid climate; the horizontal motion had left a kink in the course of the stream, creating a small dog-leg in the channel. I greedily measured everything I could find, finally showing that the eastern side of the fault had moved a few meters both downward and southward, with about twice as much downward as southward motion. Much of this may have occurred during the last big earthquake. I felt rather pleased with myself.

I was also interested in how fast, geologically speaking, the fault was moving. There was a clue to this at one place along the coast where the fault line crossed over the coastal hills and intersected the coastline, heading out to sea. Here, there is a marked coastal bench or terrace that extends along the shoreline, positioned several meters above the high-tide level. This bench is covered by beds of beautiful scallop seashells; when they were living, the scallops would certainly have been underwater at low tide. The shells I examined were still fairly lightweight and thin, a good indication that they were not ancient fossilized shells. I guessed they could be no more than a few thousand years old. There are only two possible explanations for their present position above the high-tide mark, up on the coastal bench: there has been either a drop in sea level or a rise of the land level. Now, it is known from studies in many parts of the world that sea level has hardly changed since about six thousand years ago, and prior to that it was much lower. Therefore, the land along this part of the coast of Chile must have been uplifted several meters in the last few thousand years. Looking more closely at the shelly bench, where the Atacama Fault cuts across it, I could see that the difference in height on either side of the fault was much less than a meter—but it was hard to measure this exactly without a surveyor's level.

Putting together my estimate of the age of the shells on the bench with the observation that the bench was not displaced much across the fault, I seemed to be able to show that the Atacama Fault was slipping, both vertically and—from my earlier observations inland—horizontally, at only a relatively slow geological rate—not more than a meter or so every thousand years. In other words, it certainly did not play the major role in earth movements that geologists had long thought, given its prominence as a feature on the Earth's surface, clearly visible from space; it had even been speculated that it was a structure similar to the great San Andreas Fault in California, slipping over thirty meters every thousand years—a truly gigantic fault. But, to explain the high Andes farther east, I was forced to conclude that the movements along the Atacama Fault were being counteracted by a much larger uplift of the whole region. Such is the power of field geology.

Thrilled at the success of my simple observations in making some sense of the Atacama Fault, I headed back to Antofagasta. The time had come for me to return to Oxford. However, I felt that I had come a long way since those nervous moments when I first landed in La Paz several weeks ago. I certainly had to come back.

✢ CHAPTER FOUR ✢

Jeeps, Motorbikes, and Other Things

> **Jeep,** *n.* Small sturdy motor vehicle with four-wheel drive (orig. U.S. forces G.P. = general purposes). (*OED*)

BACK IN OXFORD with my Bolivian convenio, I felt rather like a politician who has just returned from a sensitive diplomatic mission with a signed agreement. The convenio, I hoped, would guarantee the future of our South American project. In early 1990 I decided to flex its muscles at the Bolivian Consulate in London. The consulate was located in the basement of a large white Georgian townhouse, accessed from outside by stairs leading down beneath the pavement—a bit like entering a Bolivian silver mine, I thought. The basement felt damp, and a number of offices off the main corridor looked empty. One derelict office labeled COMIBOL—an acronym for the Bolivian State Mining Company—seemed to be in more or less the same condition as the Bolivian mining industry itself. At the end of the underground corridor was a room with a half-open door marked "Consulate." A tentative knock brought me into contact with Luis, the consul's son, who was to play an important role in helping us get established in Bolivia.

The consul was out and Luis was unofficially holding the fort. He spoke perfect English, with an accent that it would be hard to detect was anything but English—this, I subsequently learned, was the product of an English public school education. I began to explain about our project and long field seasons. Luis looked serious: "We would need to set up official links with a Bolivian organization, or perhaps the government itself, before such things could even be contemplated." I had a vision of high-level negotiations between the British prime minister and the Bolivian president, in which the vexed question of the Oxford University project was discussed along with military and trade issues.

I tentatively offered him our convenio. "You have a convenio?" Luis asked incredulously. He looked through the various clauses of the convenio while I sat on tenterhooks, wondering whether this was going to be enough. Luis relaxed. He pronounced that there would probably be no problem about visas—these might even be issued free of charge. He strongly recommended that we set up our base in Cochabamba. He himself was just about to go back to Bolivia and would be living in Cochabamba. Of course, we would need to rent a house and buy a vehicle. He had friends who could help us with that. If I gave him details of what we wanted, he might even be able to arrange these before we arrived. And so Luis became our official agent in Bolivia, making introductions and advising us. This was a role, I think, that Luis relished.

The Royal Society

Despite my enthusiasm for the South American project, there was something worrying me. Though our oil company sponsors had agreed in principle to support us for five years, they retained the right to withdraw at the end of each year. Oil companies are notoriously fickle when it comes to funding. When the price of oil is high, they will pour money into research and exploration. However, if the price of oil subsequently drops, research and exploration are usually the first victims of the ensuing savage cost cuttings. An advertisement by the Royal Society of London for a ten-year research fellowship, offering both a salary and generous funding, caught my eye. I knew that there would be enormous competition for these fellowships, with hundreds of applicants for only a few places. But it could secure my future for the next ten years.

I examined the application form, weighing my chances. I could think of several scientists whom I could ask to support me. Would that and my past record be enough? I decided that the chances were small, but it was worth trying. So I filled in the form and sent if off. Thereafter, the arrival of any letter would set my heart pounding. Eventually, several months later, I received a package with the Royal

Society crest boldly printed in red on the outside. The first sentence of the enclosed letter announced that the Royal Society was pleased to offer me a research fellowship. The remaining pages were concerned with the terms and conditions of the fellowship. I still find it hard to understand why I had been selected out of all those applicants. However, the Royal Society fellowship secured the future of our South American project. We had more time and money for field work. And we would be in a better position to weather the storms created by the ups and downs of the oil industry.

Back to Bolivia

Our first major field campaign in Bolivia began in April 1990. On this second trip I was accompanied by our new student, Lorcan Kennan, and fellow geologist Leonore Hoke. Lorcan was a bearded Irishman who had recently graduated from Dublin University with glowing reports. He had just the right combination, I felt, of scientific curiosity, sociability, and a relaxed attitude to life to cope with Bolivia. Leonore was Austrian and had spent many summers in the high parts of Austria's Eastern Alps visiting rocky outcrops that were usually the preserve of mountain goats. She had an optimistic outlook and did not worry too much about possible problems until they were upon her, and even then she was not too fussed — this left vacant a role that I unwittingly stepped right into, that of resident worrier. But I suppose somebody has to be this. We arrived in La Paz with the feeling of being on a true expedition. Our equipment, packed in blue plastic barrels unexpectedly donated to us by a plastics manufacturer in Britain, had been sent by airfreight. We soon made contact with Luis, who had arrived only a few days before us. Luis enthusiastically took on the job of helping us, and to this day I feel very much indebted to him.

The first task was to clear our equipment through customs. The Bolivian Geological Survey had assigned somebody to act as a guarantor that the equipment we brought into Bolivia would ultimately be taken out of the country. However, I had been advised that we should, on no account, let the equipment be transferred to the bonded warehouse. Getting something out of this was reputably more difficult than breaking into Fort Knox and would certainly involve a lot of paperwork. The best chance of getting our equipment into the country quickly was to be at the airport the moment the plane landed and persuade the airline to hand it over immediately.

Finally, the day came when Bolivian Airlines—LAB—confirmed that our consignment would be landing. I contacted both GEOBOL and Luis, and we raced up to EL Alto just as the blue barrels, strapped to a pallet, were being unloaded with a fork-lift truck. Half an hour later we were on our way back into La Paz with the barrels, conspicuously labeled Oxford University Andean Expedition, tied to the roof of a taxi. The following year I realized how lucky we had been to get LAB to hand over the consignment; this time the luggage went straight into the bonded warehouse and took thirty-six signatures on a multipage form—several of the signatures from the same official—before we could get the items released.

In the hotel we checked that everything in the blue barrels was intact. I had brought a gasoline engine drill designed for coring cylinders of rock. I would be analyzing them in Oxford to determine the magnetic properties of the rock. Leonore had packed special equipment for sampling gases and water emitted in volcanic fumaroles and hot springs. She was particularly interested in sampling helium gas, a gas so mobile that it can diffuse through most materials. We had also brought an assortment of sledge-hammers and heavy duty plastic bags for collecting rock samples.

Our Base in Cochabamba

Luis had convinced us that we should establish our main Bolivian base in Cochabamba. His ex-girlfriend's mother had an apartment she was trying to rent. Though the rent was high by Bolivian standards—four hundred U.S. dollars a month—it seemed a good deal at the time. The apartment was at the junction of two streets in a residential part of Cochabamba. Unfortunately it was completely unfurnished, so we had to more or less camp in the large and empty parquet-floored rooms. Each morning we were woken by the strange sound of somebody shouting "Tiempos, Tieeeempos, Tieeeeeeempos." I eventually discovered that this was a local Cochabamba newspaper. The other regular sound was the screeching of breaks and the inevitable crash as two cars collided at the street junction.

We needed a four-wheel-drive car. I had settled on a Toyota Land Cruiser—there were a large number of these in the country. We reckoned that most Bolivian mechanics would know how to repair one and there would be a good distribution of spare parts. Our budget would only stretch to a secondhand vehicle. The question remained whether we could get one in sufficiently good condition that we could rely on it in the remote parts of the Andes. Luis advised us that if any potential seller got wind of the fact that we were foreigners—*gringos*—the price would rocket sky high. So it was agreed that he would act as a front man, seeking to purchase a vehicle on behalf of another Bolivian friend. For this reason, we were very much in Luis's hands, unable to see or assess any potential vehicle. Each day Luis would tell us how he had seen the perfect vehicle, how he had pleaded with the owners to sell it at a price within our budget—less than ten thousand dollars—and how they would not come down. It was demoralizing as the days turned into weeks, and we were still stuck in Cochabamba without our own transport. I used to look longingly at any Land Cruiser passing down the street, thinking how incredibly lucky the owner must be.

One day Luis turned up beaming. He had found just the vehicle for us. It was owned by a doctor who was presently working in Venezuela, and it had been stored for several years on blocks in a garage. It would cost US$9,800, just within our budget. Luis's cousin was president of the Cochabamba branch of the Bolivian Automobile Association. He had examined the vehicle and pronounced it sound. And so, the next day, I found myself in one of the many *cambios*—currency exchanges—in the main plaza of Cochabamba, changing ten thousand dollars worth of traveler's cheques. Afterward I crossed over the road to the plaza, dodging the traffic, and headed for Luis, seated on a park bench in the shade of a palm tree. My bag was stuffed with dollar notes. I looked around nervously, expecting to be attacked at any moment. Luis received the notes, cleverly using a large newspaper to shield us from inquisitive eyes. He stuffed the notes into his voluminous pockets and disappeared into the traffic. The next day, a recently groomed beige Toyota Land Cruiser was parked in the courtyard of the Bolivian Automobile Association waiting for its new owners. Over the next few years that vehicle served us well, and despite numerous mechanical problems, it never broke down to the extent that we could not get it going again.

I had one more important task before we could contemplate leaving Cochabamba. I had to help Lorcan get started on his field work. I had decided that the best way for Lorcan to travel around was on a motorbike; it had many advantages over a car. For a start, it was much cheaper. Also, off the main road I felt a motorbike was much safer. The worse that can happen is that you fall over. An off-road car accident is potentially much more serious. I had nearly been killed during my thesis research in Swaziland when the vehicle I was driving came off a narrow and steep track, crashing down over a hundred meters. As a result I had switched to a motorbike for field transport, something that was, alas, not feasible for our main work in Bolivia.

Lorcan was looking forward to riding his motorbike through the Andes. We had decided that a cross-country trail bike, which was strongly built with powerful gears, was the best option. The bike we selected had an aggressive look, with wide, knobbly tires and huge springs on the front forks. Lorcan, when seated on this machine, was transformed into a different person. When he tried it out for the first time on the streets near our apartment, he became an instant sensation. Local girls looked at this young foreigner with new ideas. Lorcan had arrived.

Werner Gutentag and Señora Ahlfeld

If there were German-speaking people about, Leonore was sure to find them. So it was no surprise when she announced one day that she had met a Bolivian of German origin—Werner Gutentag—who owned a bookshop in Cochabamba called Los Amigos de los Libros. Werner had told her that his family had come out by boat to South America in the 1930s, first to Panama and then on to Bolivia, escaping Nazi Germany. Bolivia was one of the few countries that was prepared to accept refugees, and Werner soon adopted this small Andean country as his new home, and ever afterward felt enormous gratitude toward Bolivia and Bolivians for helping his family in their hour of need. He was a very cultured man who not only sold a wide range of books on South American art, history, and science, but also encouraged new authors and published their works. He was intrigued by our project and immediately suggested that we meet Señora Ahlfeld, the widow of Federico Ahlfeld, Bolivia's greatest geologist, who was also a refugee from Europe. Federico had written the only authoritative work—now completely unobtainable—on the geology of Bolivia, summarizing all that was known about the mineral deposits of this part of the Andes. He had died in the 1970s, but his widow still lived in Cochabamba.

We soon heard that Señora Ahlfeld would be very pleased to meet us and had, in fact, invited us all to tea. So, one afternoon, Leonore, Lorcan, and I, together with Werner, found ourselves in Señora Ahlfeld's garden in a quiet part of Cochabamba. She was a small and elegant—almost birdlike—woman, in her eighties, who was determined that we should be properly entertained for tea. On a table covered in a spotless white cloth she had arranged plates of sandwiches, cakes, and biscuits, serving tea from a silver teapot. She could speak English, German, and Spanish, though her English was a little rusty. Nonetheless, it was a matter of pride for her that she spoke to Werner and Leonore in German, and Lorcan and myself in English. "It is only proper that every person is addressed in their own language," she said. I smarted with embarrassment when I thought about my terrible Spanish. Leonore, at least, spoke English and German fluently, and Lorcan was rapidly picking up Spanish.

After an hour or so, Werner decided that Señora Ahlfeld was getting tired and indicated that we should go. She was determined that we stay—I think it had been a long time since she had talked to geologists, and we had stirred up many old memories. When we finally came to take our leave, she pressed on us a number of Federico's geological papers, including his book on Bolivian mineral deposits. She had clearly decided that our Andean project was the best home for them, and they are now a valuable part of the Andean library in Oxford. These gifts must have prompted Werner, because he suddenly let on that he had a large collection of geological maps of Bolivia, made by the Bolivian Geological Survey under Federico's guidance in the 1960s. The maps have long been out of print, and even the Geological Survey does not have copies of many of them. When we came to see Werner's collection we were stunned—he had a scientific treasure house of geological information. I could hardly believe my ears when Werner said that he wanted to give us the maps. We eagerly examined them, seeing the bedrock of the Andes laid out in front of us as a patchwork quilt of color—each color was a rock formation laid down in Bolivia's remote geological past. To us, well versed in the language of this type of map, they were a Rosetta stone, saving us months, if not years, of work, directing us right to the place where we should look.

Over the years, whenever we were in Cochabamba, we would call in on Werner and tell him what we were up to. One year he told us the sad news that Señora Ahlfeld had died. And Werner himself was not too well. But he had opened a small window for us on a Bolivia that is now lost forever, and I always think of both him and Señora Ahlfeld with a strong feeling of affection.

On the Road

Soon, like fledgling birds, we felt confident enough to try leaving our "nest" in Cochabamba and venture out on the open road. The Bolivian road network is heavily policed. A barrier or *tranja*, as it is called in Bolivia, blocks the road at the entrances and exits to most villages and towns. It is necessary to pay at a small booth close to the barrier, providing, in addition, information about your vehicle and journey before being allowed to proceed. Over the years, I must have passed through hundreds of these barriers. Sometimes the policeman would take my driver's license or passport and try to get the information for himself. This usually resulted in some interesting entries in his log book. Once, one of our group was recorded as Señorita Green Eyes. I was often Señor Henry—my middle name. I never discovered what they did with all this information on every vehicle that passed through. It must have been a relict from the period when Bolivia was ruled by endless short-lived and dictatorial military juntas, and the army wanted to keep control of all movements of people within the country. By the 1990s I suspect the tranjas had merely become a way of providing employment. Some tranjas were friendly; they would get to know us and often, on a hot afternoon, just wave us through. Others could be very intimidating, demanding our passports or convenio.

Our first journey in the Land Cruiser was up a rough road that led to Inca ruins overlooking the valley of Cochabamba. This was also a region where our newly acquired geological maps suggested that we would find sedimentary rocks, laid down during the last few tens of millions of years when we believed the mountain ranges themselves were rising. Lorcan was planning to examine these rocks as part of his thesis work, making a more detailed geological map. This was the first opportunity we had had to try the Land Cruiser out in off-road conditions. At first it seemed to make good progress, pulling powerfully up the steep track. However, as we climbed higher, the engine would occasionally miss a beat. This became more frequent, and eventually the engine stalled. It just did not seem to have the power to continue farther up the road. Disappointed, we managed to turn around, free-wheeling downhill for a short distance before the engine kicked into life again.

It turned out that the Automobile Association specialized more in cleaning cars than in repairing them—Luis's cousin recommended a mechanic in the backstreets of Cochabamba called Jesus. Jesus listened to our tale of woe and immediately started dismantling the fiendishly complicated carburetor on our engine. He extracted the two jets and blew through them, exclaiming that they had been put in the wrong way around. It seemed that our carburetor had not been set for the high altitude. This is a perennial problem for gasoline engines in Bolivia. The engine demands a certain amount of oxygen before the gas will ignite—the gas/air mixture is adjusted in the carburetor. In Bolivia, at altitudes over about 3,000 meters, gasoline engines set for sea-level conditions have virtually no power—there is simply not enough oxygen in the air to mix with the gas, and the engine tends to flood. This problem can be solved by setting the carburetor for a leaner mixture of gasoline. It seemed that the reason we lost power on the road to the Inca ruins was that we had driven above 3,000 meters. Jesus adjusted the carburetor and the problem vanished.

The time had come for us to start work in earnest on our research. We had arrived in Bolivia in April, and now it was approaching the end of May. It had taken over a month to get ourselves ready. We had before us the Southern Hemisphere winter, from May to October, when the Bolivian Andes are dry and hot during the day and freezing cold at night. This is the dry season when rivers are low and dirt roads are hard—the ideal time to explore the mountains. By November, when the rains begin, the weather becomes less predictable and the roads can rapidly deteriorate into greasy and all-engulfing mud.

✣ ✣ ✣ ✣ ✣

Finally we were on our way, free to investigate the high Andes, not only in Bolivia, but in northern Chile and Argentina as well. We had a few months left to us before the call of Oxford became too strong. But over the next ten years we would come back again and again, sometimes staying only a few weeks, sometimes many months. I saw myself as the common thread that ran through all of these trips, trying to follow a steady line of research, with Leonore Hoke and Lorcan Kennan working alongside on their respective projects. And I was always keen to invite other geologists to join us; they gave us a fresh view of what was becoming a very familiar landscape.

I am now going to take you with me, in the rest of this book, on a series of scientific excursions—geological expeditions—into the Andes, exploring the main strands of our research and showing the way we thought about these mountains. As we go, you will see what it is like to work as a geologist in the field, trying to make sense of the rocks. In my mind, both the scientific quest and our day-to-day existence in the Andes were inextricably interwoven. It was only because of the science that we were here at all. But the success of the research depended on our ability to cope with the many unexpected events that seem to be part and parcel of life in Bolivia.

I will begin—in the second part of this book—by showing how we can use simple geological observations to prove that the Andes are not fixed and permanent features of the planet but have grown, almost like a living organism, as they have risen above the surface of the Earth. These observations have made it possible to visualize their growth, rather like watching a film, from the moment of birth to mature adulthood. I will show how we can discover what has happened, not only above ground level, but much deeper down, inside the Earth, as the whole western side of South America has changed shape.

In the third and final part of the book, I will go behind the scenes, as it were, of this film, trying to work out how it was done. There are links between the creation of mountain ranges and the fundamental behavior of the Earth's interior. But I believe that there are also many surprising links with the atmosphere, oceans, and living organisms, demonstrating the profound interconnectedness of the workings of our planet. All this comes from geologists' attempts to answer the simple question I originally asked: why are there high mountain ranges on the face of the Earth, in the continents?

The 6,439-meter peak of Illimani looks down on noisy La Paz, Bolivia's main city. The affluent suburbs have spread downward, whereas the poorer parts climb up the hillsides to 4,000 meters above sea level.

Death at high altitude — a small cemetery on the road to the Zongo Valley in the Cordillera Real, just north of La Paz. The snow on the 6,088-meter peak of Huayna Potosi hides a bedrock of once molten granite, intruded into the crust about 220 million years ago.

The main street of Corocoro, once the location of the world's largest native copper mine, but today down on its luck.

Traveling with the Bolivian Geological Survey on my first trip to Bolivia in 1989. Raul Carrasco, on the left, was my geological guide, with our driver. The white container on the roof rack is full of a local brew called *chicha*.

Leonore looks out over the valley of Cochabamba — a rift in the mountains at an elevation of 2,500 meters above sea level.

Washing day at a hot spring in the Altiplano, near Oruro.

Water, water everywhere, but not a drop to drink—the Pacific Ocean meets the driest coast on Earth, in the Atacama Desert of northern Chile on the western edge of the Andes.

Me somewhere in the high Bolivian Andes.

Lunch break in the Eastern Cordillera: Lorcan offers a tea bag to a Bolivian sheep.

Looking across the clear blue waters of Lake Titicaca, at 3,800 meters above sea level, toward the high peaks of the Cordillera Real.

PART TWO

✣

✣ CHAPTER FIVE ✣

Looking for the Source of Ancient Rivers

Source, ***n.*** Spring, fountain-head from which stream issues.

River, ***n.*** Copious stream of water flowing in channel to sea or lake or marsh or another river. (*OED*)

AT THE BEGINNING of each trip to Bolivia I always start a new field notebook — a heavy-duty surveyor's notebook made with high-quality rag paper, sewn and bound in a distinctive yellow cover. Over the years I have acquired quite a stack of these; their front covers are labeled in large numbers with the field season's dates. My notebook for 1991 shows that I had decided to tackle, head on, a geological phenomenon I had noticed on my first reconnaissance trip to the Bolivian Andes: vast quantities of sandstone — often with a distinctive red color — in the bedrock of the Altiplano. I had found an enormously thick sequence of these sandstones — some geologists might say outrageously thick — around the copper mine of Corocoro.

Letting the pages of my notebook fall open more or less at random, I find that on August 11, 1991 — my birthday, in fact — I had noted down at a locality in the Altiplano the following: "medium-to coarse-grained fluvial red sandstones west of the village of Corque. Strata dip steeply 50 to 60 degrees west. Abundant tens-of-centimetre scale cross-bedding, with palaeocurrents generally westward. Sample of biotite-bearing volcanic ash collected at locality 3 for dating." I had also made a detailed sketch covered in geological measurements. You might well ask what all this note-taking was about. To me, it was already clear that the rocks were laid down by countless ancient rivers that once ran through the Andes; the notes were an attempt to extract as much information as possible — in the shorthand of geological jargon — about what these rivers were like, and where and when they flowed.

My view of rivers is that of a geologist, a view that is subtly different, perhaps, from the way many people see them. I see rivers as long and sinuous watery conveyor belts, transporting rock fragments from the highlands to the lowlands. Inextricably linked to this idea are two geological processes: at various points along the watery conveyor belt, where the rock fragments first join it, the underlying bedrock is being progressively eroded, while at its end the bedrock is being built up by the accumulation of this detritus. Imagine a slab of rock perched on the edge of a cliff face, high up on the side of a snow-capped mountain; it broke off a rocky outcrop and tumbled down from even higher up the slopes, coming to rest here many years ago. There is a distant rumble and the mountainside begins to shake—you are witnessing another earthquake. In all the confusion, the slab slowly slips forward, teetering on the edge before falling down to the boiling river, far below. It hits a huge boulder in the river and smashes into many pieces. Each fragment is swallowed up by the water's turbulent flow, and the mountain has finally lost a small part of itself. Carried along in the current, the fragments are hurled against each other, pulverized into grains of sand, silt, and mud. Over time, many more rocky bits of the mountain will be consumed by the greedy river.

The river will take its cargo of grains on a long and violent journey down to the lowlands. In the process it will undergo a profound change itself. First, it is a torrent of water cascading over boulders and down steep and narrow gorges. Next, the river slows and snakes through the mountains across a wide bed of shingle. Finally, it spills out onto the sandy plains of the lowlands, meandering through thick vegetation. The volume of water in the river will fluctuate with the seasons. At the height of summer, when no rain has fallen on the hills for many months, the river slows to a trickle, sluggishly finding a route through huge gravel and sand banks. In the spring, however, when the winter snow starts to melt and after days of heavy rain, the river fills its banks with muddy brown water.

Once in every hundred years or so, the spring rains will be unusually heavy. The river will rise up from its bed like an enraged monster. The energy of this huge volume of water, propeled by gravity as it cascades off the mountains, is enormous; enough to transport millions of tons of rock fragments. Once in the lowlands, the river loses much of its power and can no longer carry its load; vast quantities of sand and silt fall to the bottom, adding a new layer of sediment to the landscape—a new page in the geological record. When the water has subsided, plants will begin to colonize the surface. These will have time enough to claim back the river banks before the next storm turns over yet another leaf of the record as it brings down more sand from the highlands. It is these rock pages that I was, in effect, transcribing into my notebook.

The rock pages reveal much more than just the existence of a river in the past. They also tell a geologist that there must have been highlands around when this river flowed. This way, the ancient river sediments can be a tell-tale sign of a past mountain range, far back in the Earth's history. In this chapter I will show you how the record of ancient rivers in the Andes can tell us much more, yielding up a remarkable story of the evolution of a landscape, going right back to the moment when the Andes first appeared as a recognizable feature on the Earth's surface.

The Creature of the Lake

It is difficult to make sense of the rivers of the past if you have no feel for how the present rivers flow in the landscape, and how they might both mold this landscape as well as be controlled by it. Where better to look at today's rivers than the Andes themselves? These mountains, with their predominant north-south axis, divide the main drainage between rivers flowing eastward and those flowing westward. I never cease to be fascinated by the thought that I can dip my hands into any river flowing off the eastern flanks of the Andes and touch water that, after following a tortuous route across the grain of the mountains, will eventually traverse the entire continent to mingle with the Atlantic Ocean, several thousand kilometers downstream. As it wiggles away from the mountain front, like a silvery snake, some of this water will join the headwaters of the

Amazon River in Brazil, or it will flow into the Parana River, reaching the Atlantic much farther south at Buenos Aires, in Argentina. On the western flanks of the Andes, the rivers—known locally as *quebradas*—have a much shorter journey, flowing straight into the Pacific Ocean. However, because almost all the rain falls farther east, the rivers here are much smaller than those on the Atlantic side and are generally highly seasonal with long dry periods, when the quebradas are just a trail of boulders.

You would be forgiven for thinking that these rivers—though perhaps changing with the seasons—have continued from year to year flowing down to the lowlands, and on to the sea, forming permanent features of the landscape. But, in fact, the rivers of today are not the same as those of the past. From a geological perspective, rivers have short lives. Ancient rivers will die, but others will be born, following new courses through the mountains; they are all at the mercy of deeper forces in the Earth. We can see this in the story of the rivers and lakes in the Altiplano of Peru and Bolivia, learning, at the same time, about something even bigger—the constant state of geological unrest of the whole mountain range.

In some ways, it is the Altiplano that makes the Andes such an unusual part of the world, forming a landscape of rolling plains and low hills right in the heart of the mountains, plains where the Andean farmers have grazed their llamas for millennia. But don't be fooled by the gentle topography—this region is nearly 4,000 meters above sea level. And despite its height, if you stand on the Altiplano, you will see in the far distance even higher snow-capped peaks. You are, in fact, on the floor of a broad and long basin, hemmed in by two parallel ranges, to the east and west. Rivers that flow into this basin cannot escape, and they end up in either evaporating salt pans or lakes. The largest and most famous of these lakes is Lake Titicaca.

I had seen Lake Titicaca on my very first visit to Bolivia, when I was trying to fill in time before meeting up with the Bolivian geologist Raul Carrasco. A local company ran hydrofoil trips to the town of Copacabana, nestled in a bay on the lake's southwestern corner. Copacabana was famous for its Franciscan monastery, home of the so-called black virgin, a dark, meter-high wooden figure representing the Virgin Mary. People from all over South America made the pilgrimage across the lake both to see the black virgin and to receive a blessing from the monks for any newly acquired possession in their lives—cars, houses, furniture, or new-found wealth—hoping that this would bring more good fortune in the future.

The special religious significance of Copacabana—the Incas also built a temple here—may really be recognition of the importance of the lake to the peoples who have lived on its margins for at least the last five thousand years. This wide and deep blue body of water, so unexpected in a barren and arid region at an altitude of 3,810 meters, has provided an abundant supply of life-sustaining fish, fertilized the adjacent plains with nutrient-rich muds during the wet seasons, and been a source of irrigation. To a geologist, though, the lake is something of an enigma. It is remarkably deep in places, reaching depths over 250 meters—you would have to sail well over a hundred kilometers out to sea before you encountered ocean water this deep. As the hydrofoil motored out onto the broad expanse of water, it struck me that there was something else odd about the lake. Its complex shoreline, with numerous bays and islands, had the classic appearance of a drowned or flooded landscape—a landscape of valleys that were once sculpted by a large and powerful river with its many tributaries. So my first journey across the lake had already begun with a geological question about a river: where did this ancient river flow from or to, and why is the region now a lake? The source of probably the most sluggish river in the world, which flows southward along the length of the Altiplano—the Rio Desaguadero—provides, I believe, a clue.

I have heard Bolivians argue about whether Lake Titicaca is the source of the Desaguadero. The answer depends on when you look. The true Desaguadero rises in a swamp some kilometers south of the lake. Several other streams flow into the swamp, including a channel that connects with the lake, following a short section of the border between Peru and Bolivia. In very dry weather, when the lake level is low, this channel is reduced to a stagnant body of water. But the channel was clearly not like this in 1539, when

Hernando Pizarro and his band of conquistadors fought a crucial battle on its banks during the Spanish conquest of the Inca Empire. The Incas had decided to make a stand here, trying to stop the southward progress of the Spanish. To begin with, things went badly for Pizarro. He was unable to reach the Incas because the channel was in the way, and eight of his men drowned in their heavy sixteenth-century armor when they tried to cross it on horseback and sank into deep water. The next day, Pizarro tried a new stratagem. He managed to transport his men and horses on rafts down the channel from their base on the shore of Lake Titicaca. This, too, nearly ended in disaster when his men lost control of the rafts in the strong current and were swept downstream. Some quick thinking saved the situation, and once his horsemen had disembarked and were mounted, the battle was won. I find this story interesting because it suggests to me that Pizarro had arrived on the shores of Lake Titicaca when the water level was unusually high, perhaps during a particularly wet spell of weather. The lake was now clearly acting as the source of the Desaguadero, spilling out southward down the channel with some energy. In effect, the channel was behaving like the spillway in a dam or weir, ultimately controlling the lake level. And when the lake level eventually dropped again, the spillway would dry up.

Even at its lowest recorded levels, Lake Titicaca, especially at its southern end, still looks as though it has flooded an extensive landscape of valleys and hills. I had reached this conclusion on my very first visit, and I am sure that anybody else who has studied geography at school would have done the same. But the valley bottoms of this drowned landscape, now deep under water, are hundreds of meters lower than the rest of the Altiplano. This is stranger than you might, at first, think. The only force of erosion capable of excavating such a depression would be a river. But a river could do this only if it could somehow flow out of it, carrying the excavated bedrock away. But how can any river flow out of what is effectively a hole in the Earth's surface? The Desaguadero today, as we have seen, is really just an overflow for the lake, and if the lake bed were dry, rain water would flow back into the depression.

Here we have a true conundrum. But perhaps we have overlooked something? What if the Desaguadero spillway was once much deeper, forming a narrow gorge? Then, the river could flow out from the bottom of Lake Titicaca. There is no gorge now, so perhaps it has been completely filled in and hidden? There are many problems with this idea. To begin with, this gorge would have to be extremely deep—a chasm cutting down several hundred meters to lie everywhere below the level of the present lake bed. And it could not just run into the Desaguadero, or any other river in the Altiplano farther downstream, because the beds of these rivers are too high, flowing way above the level of the gorge! So the gorge would have to be much longer, leading out of the Altiplano and linking up with one of the main Andean rivers that flow down to the sea. And finally, in order for Lake Titicaca to form, the gorge would have to be somehow blocked and filled along its entire length with detritus. It seems a tall order for such a gorge ever to have existed.

There is, in fact, a far simpler explanation. But we first have to give up our deeply held conviction that the bedrock of our world is solid and fixed. We can then entertain the idea that Lake Titicaca owed its origin to large vertical movements of the Earth's crust, movements that created the depression that it now fills. There must have been a time when rivers flowed out of the region down to the lowlands, sculpting the original valleys that we can clearly see in the shape of the lake. I believe that both the lake bed and surrounding landscape were extremely unstable. The peaks that now hem in the Altiplano rose up all around, blocking the path of the original rivers and trapping their waters in a natural basin. But the floor of this basin was not fixed either, but warped up into a series of swells with intervening depressions. Eventually, the deepest of these depressions filled with rain water. This is Lake Titicaca, and the Desaguadero spillway flows over the top of the swell into another, much shallower, depression farther south. As I journeyed back across the lake from Copacabana, on my first visit to Lake Titicaca, my head was full of this idea. I imagined dashing off a paper to the prestigious scientific journal *Nature* putting forth my new theory.

However, as my excitement began to cool, I realized that I would need far more evidence.

Over the years, I feel that I have found the evidence, in the process developing a grander image of the Andean mountains. These mountains are not static and dead lumps of rock, but mobile features of the planet that, over time, have risen up to their present elevation. To my mind, their rise conjures up an image of some strange creature emerging from the depths of an almost mirrorlike expanse of still water. Before the mountains existed, the creature lay far below, sound asleep. But when the creature awoke, it began to rise slowly to the surface. At first, the only signs were a few ripples disturbing the reflections on the water. Then, something solid, with more form, broke through, moving a little to one side and then to the other, and gradually rising higher and widening—the shape of the creature became clearer. Superimposed on the rising shape was a gentle undulation as some parts of its body rose a bit more and others slightly subsided. Eventually, the creature stood above the surface, revealing itself at its full height—the Andes, as they are today.

We can see in this image an analogy for the restless behavior of the rivers. A steady downpour of rain would cause rivulets of water to snake across the monster's back, seeking out the low points as they made their way downhill. Without a dam to stop it, the water would flow straight off the creature into the lowlands, forming the great rivers of the Andes that ultimately reach either the Pacific or the Atlantic Ocean. But any barrier in the way would block the flow, creating pools of water. These pools are the high Andean lakes such as Lake Titicaca. As the creature rose, the low points and barriers moved around, constantly changing both the courses of the rivers and where the water ultimately ended up. Sometimes new lakes were formed. Sometimes old lakes were emptied and became part of a system of new rivers.

You might well think that the creature-of-the-lake analogy is far-fetched, a geological Loch Ness monster, perhaps. Many people have spent an inordinate amount of time watching Loch Ness, scanning its dark waters with high-powered binoculars, or sounding its bottom with sophisticated devices, hoping to see Nessie poke its long neck out of the lake. They are wasting their time. However, watching the rise of the Andes is not a hopeless undertaking. In the rest of this chapter I will give you hard evidence, which, when put together, will enable you to witness their emergence from the deep.

Dunes in the Rocks

If you dig a trench into the sandy banks of a river on the lowland plains, east of the Andean mountains, you will find a detailed record of previous river floods, when the river was a raging torrent. Exposed in the walls of the trench are layers of sand, and each layer will contain even thinner laminations picked out by slight changes in the size and color of sand grains. These laminations are not straight. They may have distinctive wavy shapes defining a series of humps and depressions or be arranged rather like successive rows of inclined slabs of rock in a stone wall—cross-bedding in the terminology of geologists. You are looking at a section cut through a series of underwater dunes. Their shapes are the clue to a method of observing the Andean monster, and the best way to understand this is to think of the sandy deserts of Arabia—the landscape in which the Bedouin peoples live.

In a desert, turbulent gusts of wind whip grains of sand into the air and then dump them back on the ground. I am sure that the Bedouin are well aware, as they wrap themselves up against the sandstorm, that these apparently random movements of sand are creating something of immense beauty and regularity in their desert world: sand dunes. And the Bedouin can tell, just by looking at the dunes, which way the prevailing wind blows—the upwind side has a fairly gentle slope, but downwind, the sand is banked up much more steeply. In high winds, the sand streams over the dune, moving up the shallow exposed side, and reaching the crest before tumbling down the precipitous lee slope. When the wind drops, and the mist of streaming sand clears, the freshly sculpted dune reveals itself with sharp lines, its surfaces sometimes corrugated with parallel rows of

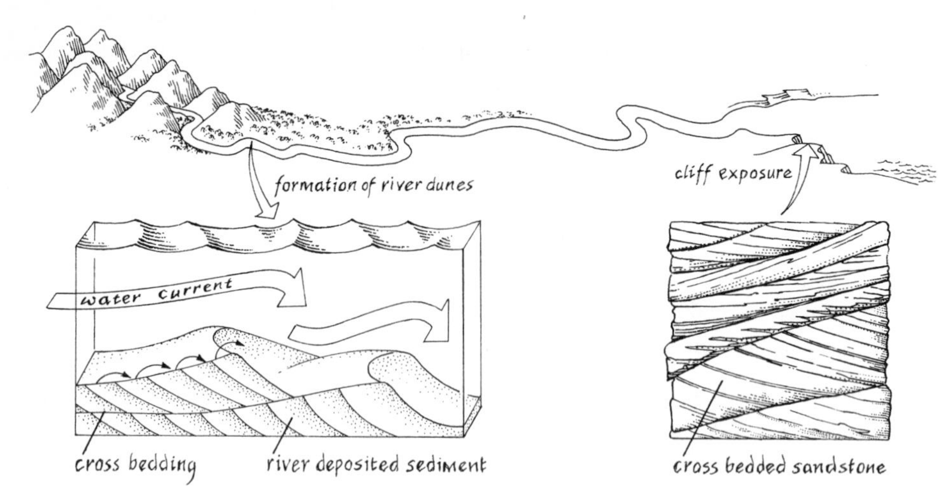

Dunes in the rocks. The sandy bed of a river is molded by the flow of water into characteristic hummocks or dunes, and the pattern of internal sand laminations shows which way the water flows. Fossilized dunes like this are the clue to a long-lost river flowing off a mountain and down to the sea.

small ripples, like little fleas on the back of a bigger flea. Inside, the movement of the sand has created thin laminations that mimic the shape of the dune.

Dunes are formed in much the same way underwater, on the bed of a river, lake, or the sea, ranging in size from much less than a tea kettle up to a small schooner, as one geologist famously remarked. Instead of the wind, it is the current of water that moves the sand along. In a river, the grains bounce along the bottom with a characteristic rolling motion called saltation, inexorably washed along by the current. During the later stages of a flood, when the river level begins to subside, more and more sand falls to the bottom, building up the surface of the dune with a stack of laminations. In time these wavy laminations will be preserved in the sand layers deposited on the river banks, and eventually, as the layers are buried more deeply, they will become features of solid rock, fossilized in sandstones.

I am always amazed by the richness of the record of fossil river dunes. Sometimes, when a rock is split open, the shapes of the ancient river dunes are exposed as they would appear if viewed from above, forming corrugations that vary in both size and shape. You are looking down on the natural features of the bed of an ancient river, molded by the flow of water and perfectly preserved for tens of millions of years. The troughs between the crests of neighboring dunes or ripples may contain even smaller ripples, formed when the shallow water was moved by a single gust of wind—a true instant in time.

Much like the desert traveler, most geologists can tell, just by looking at the shape of the river dunes, which way the water was once flowing—a direction called a paleocurrent—because the steep dune face slopes down current. And, of course, water always flows downhill, away from the highlands and toward the lowlands. This very simple idea suggests that the fossilized river dunes, preserved in the rock layers, can tell you not only which way an ancient and long lost river was flowing, but also the direction in which the highlands lie. It is a short step from here to using the dunes to work out the pattern of ancient rivers flowing off these highlands, revealing the full extent of mountain ranges at various stages in their growth. Here was a way to study the thick sequences of sandstone preserved in the Bolivian Andes. I was searching for evidence in the river layers of what the mountains were like far back in geological time, long before there were geologists, or even humans, to observe them.

The First Andean Rivers

The layers of sandstone left behind by rivers are easy to find because they usually have a distinctive red color—geologists call these sandstones red-beds. The color comes from minute particles of iron oxide or rust—the mineral, hematite—that coat the grains of sand in the rock. The color is not just on the surface. If you break off a lump with a hammer, you expose fresh rock faces that are also red—the rock is rusted through and through by contact with oxygen when the sediment was first laid down. Since the demise of the dinosaurs, layer upon layer of red-beds were laid down in many parts of the Bolivian Andes by the forerunners of the present rivers, forming vast piles of sediment, many kilometers thick. The rivers' remains are now found in the high Altiplano, the rugged Eastern Cordillera, and in the foothills of the Sub-Andes. They, together with even older rock layers, have been pushed over by Earth movements, so that tilted on end, they stick up out of the ground. This makes the job of reading the rocks much easier because as you travel over the terrain, traversing the tilted strata, you are also making a journey through geological time, moving from younger to older layers.

On many journeys through the arid landscape of the Altiplano or Eastern Cordillera, I have traveled far back in time into the older strata, eventually reaching Cretaceous-age outcrops, formed when dinosaurs thundered past. These are not red-beds, but steeply tilted slabs of limestone alternating with siltstone; layers that were once carbonate oozes and silts that settled out on the bottom of a large body of water. Sometimes this body of water formed a land-locked lake, but at other times it joined the world's oceans. Fossilized shark teeth in some of the layers testify to an occasional sea connection—

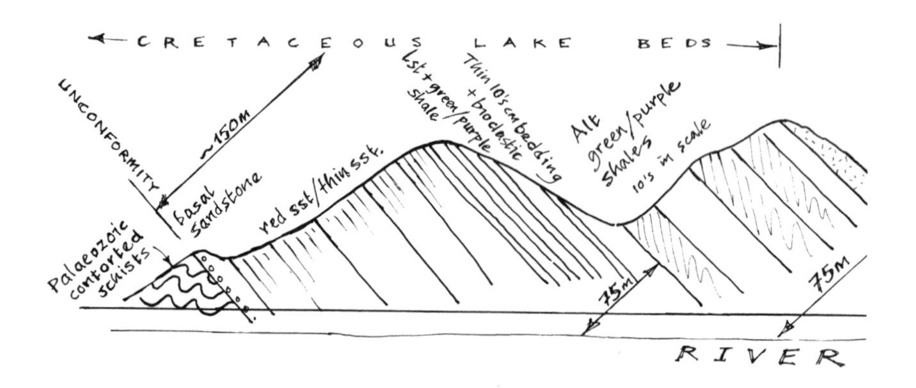

Another sketch from my field notebook. On the eastern edge of the high Bolivian Altiplano, tilted slabs of rock are clearly visible in the banks of rivers. Traversing the layers is a journey back in time to the period, at the end of the Cretaceous, when the region was a huge inland sea or lake and layers of perhaps when sea level was particularly high and overflowed the low barrier separating it from the lake. But when water levels were lower, during generally drier periods in the Cretaceous, the inland sea became a series of large, isolated lakes in which brackish and freshwater fish, now fossilized in the rocks, thrived. Sometimes the lakes dried out altogether, leaving a layer of salt.

I once showed the limestone remains of this Cretaceous sea, exposed near the town of Camargo in southern Bolivia, to an Argentine geologist—Eduardo Rosello. The rocks were full of minute spherical balls, called ooiliths, that form in shallow tropical waters, like the Bahamas, when small grains of limey ooze roll around on the sea floor, moved by variable currents. Eduardo assured me that he had seen exactly the same features, in rocks of the same age, exposed in northern Argentina, hundreds of kilometers farther south—the only difference was that they call these rocks the Yacoraite Formation, whereas the Bolivian geologists have given them the name El Molino Formation, after the place near Potosi where they are best exposed. Once we had made sense of this confusion of names, the truly vast size of the Cretaceous lake or inland sea started to become clear. But by about midday, after a morning of scrambling over sun-scorched rocky slopes in the broiling sun,

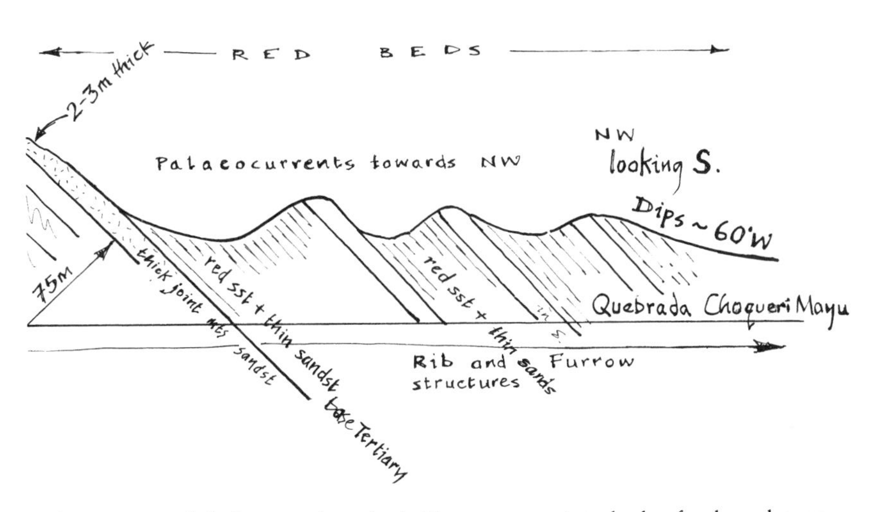

limestone and shale were deposited. These pass up into beds of red sandstone, laid down when the present mountains first started to rise and rivers flowed across the landscape.

we had had enough of limestones, whatever their significance—the back of my shirt was soaked with sweat, and the corners of my eyes were aching from squinting in the bright light reflected off the pale-colored rocks.

Eduardo proposed a break. He had noticed that the lower slopes of the hillside were cloaked in vines. We soon found the center of this wine industry—an old hacienda where the grapes were being distilled into a clear spirit called Singani, though the owners kept back some of the harvest for true wine making. The vineyard showed us unexpected hospitality, serving ice-chilled white wine in tall glasses as we reclined in hammocks on the deep-shaded veranda of the old residence. Vines curled up the supporting pillars, and spread out in front of us was a view of craggy hillsides of green growth, made vivid by the underlying bright orange calcareous soil. But this idyll could not last, and we were soon back in our baking hot Land Cruiser, pounding down yet another dusty road in Bolivia, tracing more of the limestone layers through the Eastern Cordillera.

There are places where sandstones and siltstone beds, interleaved with the limestones, are marked by dinosaur footprints—trails of small pits that traverse the exposed surfaces of the rock layers. Near the village of Torotoro, these footprints are particularly easy to find; one only has to follow the paths leading to and from the village to see them. These paths, worn by generations of llama and mule trains, pass over great slabs of gently inclined bedrock, sometimes climbing up through successive layers in a giant natural staircase—a staircase in time as well as space. Beneath one's feet, the dinosaurs have left their mark, in places apparently following the modern track. They must have been thriving here in large numbers because it is a curious fact that more of their footprints have been found in the rocks of the high Andes of Bolivia than anywhere else in the world.

Herds of titanosaurs, lumbering side by side, or fleet-footed carnivorous dinosaurs, crossed mud flats or pools of shallow water at the edge of the inland sea in the summer months, leaving tracks as their feet sank into the soft ground. Later, during heavy rains, the landscape was awash with muddy water. The mud settled out, delicately filling the dinosaur footprints and eventually incorporating them into another layer of rock to be preserved for tens of millions of years. The dinosaur layers serve as a clear physical marker, easily recognizable throughout the Bolivian Andes, of a time when the whole region was close to sea level—like the flat sheet of water, perhaps, in my image of the rise of the Andean monster—a time before the mountain range existed. In reaching these layers I had traveled too far into the past, and I needed to retrace my steps back up through the rock layers, looking for signs of the mountains.

The main path out of Torotoro makes its way down to the River Caine, one of the major rivers that flows through the Bolivian Andes. The rock layers are tilted themselves toward the river, so that as one follows the path, one is slowly progressing up through the strata, moving forward in time. Eventually, the dinosaur footprints are left behind and one reaches layers of silt containing the bones of small rodents—fossilized primitive rats found by paleontologists from the University of Salta in identical layers in northernmost Argentina. Here, no sign of the dinosaurs can be found, strongly suggesting that these layers were laid down after the dinosaurs had gone extinct, in periods of time called the Paleocene and Eocene by geologists—the beginning of the age of mammals. Continuing on, the path leaves the silt layers and leads into a narrow gorge, cutting through a high ridge of rock before eventually coming out on the bed of the River Caine.

Exposed in the sides of the gorge are layer upon layer of the red-beds I was looking for—the remains of ancient rivers. These rivers must be the first indications of mountain building in the region, when the rocky bed of the Cretaceous inland sea started to rise above sea level and rain water began to flow off the newly formed hills. Much like their modern counterparts, the rivers scoured into the uplifted bedrock, carrying away detritus and depositing it at the foot of the mountains. This way, the pile of red-beds started to accumulate. And the shapes of the dunes in the layers contain the clue to the location of these lost mountains: the steep lee face slopes in the direction of river flow—the paleocurrent—away from the highlands and toward the lowlands.

On many occasions during our field work I have organized dune-hunting expeditions, reaching remote places with names that sound like a mantra in the Aymara and Quechua languages: Villoma, Sarcosa, Chanar Mayu, Potoco, Tambo Tambillo, Nekke Khota, to list a few. Our target was usually a series of rocky bluffs, like those in the gorge near the River Caine. Once there, we would fan out, scrutinizing the shapes of the ancient river dunes in the rock layers exposed in the cliffs. These are usually clear to see, picked out by either the roughness of the rock surface or delicate lines of differently colored sand grains. It is not enough, however, to look at one dune. This is because rivers—modern and ancient alike—are not perfectly straight but tend to snake their way through the landscape. So I would often set a target of ten dunes for each member of our dune-hunting team—usually they would grumble at being put to use like this. We would look at the directions in which the lee faces sloped for a whole succession of layers, laid down in variously orientated reaches of the fossilized river system. Combining all our observations, we had a good indication of the general flow of water that laid down the red-beds—a paleocurrent for a long-lost river.

The results of my study of the fossil dunes exceeded my expectations, clearly revealing the extent of the Andean mountains when they were still young features of the Earth's surface. The paleocurrents for the early red-beds on the eastern margin of the Altiplano—deposited near the base of the red-bed pile—indicated rivers flowing in a range of directions but, on average, westward. Clearly, the source of these rivers was farther to the east. Yet in the east, paleocurrents from similar sandstones, exposed in the steep mountainsides of the Eastern Cordillera, suggested rivers flowing in the other direction, from a source to the west. The only way to explain this divergence in the rivers was if they were flowing off two sides of a narrow range of mountains, not more than a few tens of kilometers wide—the scale, perhaps, of New Zealand's Southern Alps—and, now, no longer alone, because the rest of the surrounding Eastern Cordillera and Altiplano has risen up all around. It turned out from a wider search for fossil dunes that there was also another early mountain tract—a sister that lay west of its sibling, and also slightly west of the present chain of giant volcanoes in the Western Cordillera. This sister range appears to have been mainly volcanic. It was located by both the direction of the earliest rivers flowing into the western side of the Altiplano and the visible stumps of much older and deeply eroded extinct volcanoes.

At this time, near the beginning of the Cenozoic era—in the Eocene period, very roughly forty million years ago—the region that was to become the high Altiplano appears to have been a lowland river plain, close to sea level and similar to the present lowlands of eastern Bolivia, but hemmed in, both to the west and east, by two uplifting ranges of mountains. This is a far cry from the present 4,000-meter-high plateau. A modern equivalent might be the Tarim Basin in Tibet, trapped between the Tien Shan ranges in the north and the Altyn ranges farther south. Rivers flowing off these two ranges dwindle and disappear in the wide deserty plains of the Tarim. They are adding detritus to a steadily accumulating pile of red-beds, just like the early rivers in Bolivia.

A Rock Time Table

As we have seen, the dunes in the red-beds are a sort of finger post, pointing out the whereabouts of the mountains at various stages in their growth. But to be able to tell the full history, I needed a yardstick by which to measure the age of the red-beds. They contain many layers, and each layer marks a beat in geological time. As the geological clock ticked on, with monotonous regularity, a thick pile of strata accumulated. Though it is easy to measure the thickness of this pile, the clock itself is more difficult to read. This is the perennial problem that all geologists face. For the first few years of our project, unless the red-beds formed part of a continuous sequence piled on top of the old Cretaceous lake or sea bed, we could not compare or correlate them from one part of the Andes to another; we had no very precise idea of when they were actually deposited, except that it was after the dinosaurs had gone extinct, in the Cenozoic era.

Ever since the pioneers of geology started to examine closely the layers in the rocks, they realized that the layers could be ordered using the fossils preserved in them. William Smith, who made the first geological map of England and Wales at the beginning of the nineteenth century, was one of the first to do this. Smith realized that many of the layers contained fossils that were "peculiar unto themselves." Charles Darwin was soon to show in his theory of the "origin of species by natural selection" that the "peculiar" relation between fossils and the rock layers they were found in is an inevitable consequence of the evolution of life. The formation of the rock layers records a moment in geological time, and the fossils preserved in the layers are those that lived at this moment. So, if strata in any two different parts of the world contain the same fossils, then you can be sure that they were both laid down at the same time. The chances of evolution throwing up exactly the same type of organisms independently at different times are just too small.

Gradually, by putting together all the information about when these organisms lived, using the work of geologists from many parts of the world, it is has been possible to construct a consistent chronology—a biological method of ordering the rocks, though it does not yield any absolute ages. This is rather like the way archeologists have used styles of pottery, architecture, or technology to work out the relative ages of ancient civilizations. For geologists it works extremely well for layers of sediment laid down on the sea floor, where abundant organisms are fossilized in the muddy oozes. These fossils have been used to order most of the sequence of events in the younger parts of the geological record, when life had reached a stage of great diversity. The problem for geologists studying ancient rivers is that organisms are only very rarely fossilized and preserved in the turbulent and exposed landscape of a river. So the chance of finding fossils in river sediments—for example, the redbeds in Bolivia—is very small. Even if they can be found—usually impressions of leaves, reeds, or grasses or odd mammal bones—they often tie the rocks to only a vague period of geological time and certainly cannot be used for detailed comparisons of widely separated outcrops of rock. We quickly realized that this way we were not going to be able to work out the timetable for the accumulation of the red-beds. We had to be able to date the rock layers much more precisely.

The answer lay in the giant chain of volcanoes that runs along the western margin of the Altiplano, in the Western Cordillera. More precisely, it lay in the fact that the abundent lava and ash erupted from these volcanoes are relatively rich in the element potassium, and some of this potassium is radioactive, spontaneously turning into another element, argon, as it decays. The obvious thing to do was to use the well-known potassium-argon method of rock dating, a method with a distinguished track record in geology. It had played a pivotal role in unraveling the mystery of the magnetic anomalies on the sea floor, opening the door to the idea of sea floor spreading and plate tectonics. It had also illucidated some key moments in our own evolution from primates to man. But it had now been abandoned by many of our geological colleagues in favor of more elaborate techniques, capable of far more precision. Yet to us it had many advantages. To begin with, it was a relatively quick and easy way to determine the age of a large number of rocks. And Chris Rundle, who, at the British Geological Survey, ran the last laboratory in the country that specialized in this technique, was willing to help us.

We were exploiting the fact that the rate of decay of potassium to argon is so steady that it can be used as a clock—a radioactive clock. Nuclear scientists have found that the rate at which it does this, as with all types of radioactive decay, will not be altered by the temperature and pressure of the rock, or any known chemical reaction. And the parent and daughter pair, potassium and argon, are particularly suitable for dating rocks. This is because potassium is a solid that will happily reside in rocks. But argon is an inert and mobile gas, and, given a chance, it will eventually find its way from wherever it is in the Earth into the atmosphere. The gas has its best chance when a rock melts. So, for example, when lava erupts at the Earth's surface, any argon gas quickly bubbles out. When the lava finally cools sufficiently to become solid, it will be free of argon, but it will still contain potassium. At this moment the radioactive clock has been set to zero. However, some of the potassium will be radioactive potassium, and this will continue decaying to argon. This argon will now be trapped in the solid rock, and so the radioactive clock starts ticking. Over time, as the clock ticks on, the argon gas builds up. All we had to do to tell the time since the lava erupted was first to measure the amounts of potassium and argon, and then to use the fundamental rate of decay of radioactive potassium to calculate how long it would take to produce the argon.

In practice, potassium-argon dating is not quite so straightforward and requires both careful preparation of the samples and measurement in a special instrument called a mass spectrometer. Chris Rundle had built his mass spectrometer in the 1970s, and it looked like a very complex piece of plumbing that extracted and channeled argon gas from the sample into the counter of the spectrometer. It had to be coaxed into action and had developed over the years a number of peculiar quirks. Leonore, Lorcan, and myself all took turns in carrying out the analyses. When it came to do my shift, I felt a bit like Charlie Chaplin on the factory production line in the film *Modern Times*, rushing to open and close valves with a torque wrench. On a number of occasions I had to face Chris's wrath after I had overtightened an awkward nut or cracked a piece of expensive glass tubing. But eventually, between us, we managed to analyze many hundreds of samples this way.

We dated not only the lava flows, but also the eruptions of volcanic ash. This ash consists of volcanic fragments blasted into the atmosphere during explosive volcanic eruptions. The fragments form vast clouds, capable of spreading out from the volcano for hundreds of kilometers. Eventually, they fall back to Earth, raining down on the landscape and covering it with a layer of ash. Much of this ash is soon washed off the hills during heavy rains. It is carried down by rivers and deposited in the lowlands as a distinctive volcanic layer. Soon, new layers of red sandstone will be deposited on top by rivers, and the layer of volcanic ash will be buried as a distinct page in the rock record. Other volcanic eruptions will leave their own mark, fixing more of the red-beds in geological time.

The potassium-argon ages for the volcanic eruptions provide time planes within the layers making up the thick pile of red-beds—a timetable for the deposition of the sedimentary layers. The dates pinpoint events in the timetable to the nearest million years or so, outlining a history that spans about twenty-five million years, from the Oligocene to the Pleistocene periods of geological time. Like an archeologist or a historian, who no longer has to speculate about a human event that occurred in some vague period of time—for example, the Bronze or Iron Age, or Dark Ages—but can specify in years or decades when it occurred, we could now discuss with much greater confidence the story of the red-beds. Putting our dates together with the pioneering work of Jack Evernden, Alain Lavenu, and Larry Marshall from the United States and France, we could now name moments of time, such as 23, 9.5, or 5.2 million years ago, when widely recognizable layers were deposited. And we could see that some parts of our story, painstakingly worked out in the Bolivian Andes, correlated with events not just in the rest of Andes, but in other parts of the globe. Frustratingly, though, we did not seem to be able to reach back with precise dates to periods beyond twenty-five million years ago; none of the volcanic material we could find had an older age, despite an extensive record of red-beds that clearly extended much further back in time—the timetable for the deposition of these older layers was much more vague. This was a problem that the red-beds alone could not solve for us, and so we had to look elsewhere. But before I describe how this problem was solved, let me tell you some more about what the rich language of the red-beds can tell us.

Unconformities

It is clear that when mountains rise, the rock layers are faulted and tilted. Eventually, older deposits of red-beds or sometimes lake beds, laid down either in the lowlands at the foot of the mountains or in depressions within the mountains, become part of the moving crust. This is happening today on the eastern margin of the Bolivian Andes, on the fringes of the Amazon region, where the Andean foothills are being pushed over the remains of river beds, laid down only a few tens of thousands of years ago. Later, a new system of rivers may flow over them, smoothing out the land surface as the rivers cut sideways and down. Even later, as the shape of the mountains changes yet again, detritus might be deposited from this new system of rivers, or even from a new large lake created by the natural damming of the river system. The newly deposited red-beds or lake beds will rest on top of the tilted and eroded older strata, forming a marked and abrupt change in the rock layers.

Abrupt changes like this have been observed in the rock record all over the world. I have already described in chapter 2 how a Scot, James Hutton, was the first to see their full significance at what has come to be known as Hutton's Unconformity. Here, in 1788, he discovered the contact between two distinct sequences of sedimentary strata while rowing with a companion along the rocky shoreline southeast of Edinburg, where nearly horizontal red-beds in the coastal cliffs rest on steeply tilted and folded grey siltstone and sandstone strata. At the contact, exposed on the shore platform, ancient pebbles are strewn over the planed-off older rocks—gravels laid down by the fossil rivers. Hutton called this abrupt change in the rocks an unconformity. Modern dating has shown that the red-beds in Scotland are, in fact, several tens of millions of years younger than the grey tilted siltstones and sandstones. With one step one can bridge the huge periods of geological time at Hutton's Unconformity.

Early on in our field work I had discovered my own unconformity in the massive road cuttings on the tortuous route across the Eastern Cordillera to Cochabamba. Here, fossilized desert dunes, blown by a wind in early Cretaceous times, were heaped up directly on top of twisted and buckled layers of deep-sea sedimentary rocks with their trilobite fossils, laid down many hundreds of millions of years earlier during the Silurian epoch. The Silurian strata must have been tilted, raised up out of the sea, and eventually smoothed

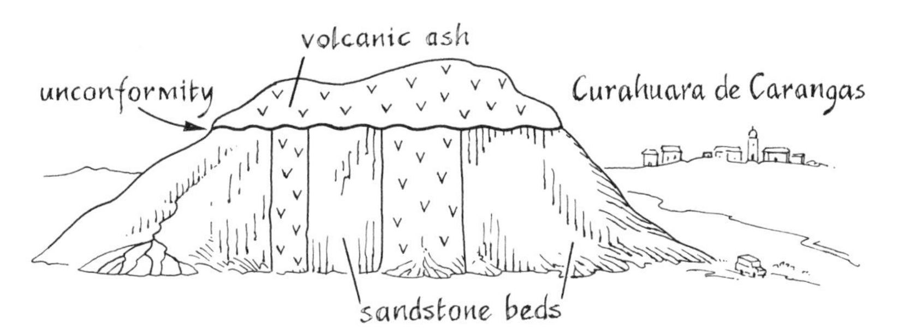

An unconformity in the Bolivian Altiplano reveals a remarkable sequence of geological events, preserving a brief moment of intense movement in the crust, about seven million years ago.

by rivers, long before the desert dunes moved across them. Unconformities like this are an extraordinarily significant feature of the rocks, pointing to a lost mountain range that existed long before the region became an inland sea or lake at the end of the Cretaceous.

Our investigation had now revealed many more unconformities, this time in the red-beds laid down after this large body of water had disappeared and the Andes themselves were rising. For example, I can turn to the pages in my field notebook that describe and sketch an unconformity, exposed in a hillside near the village of Curahuara de Carangas in the barren high Altiplano. In the flanks of the hill, layers of volcanic ash stood sandwiched between red-beds like vertical walls of white, powdery rock. We had dated the volcanic eruption that had thrown up this ash to somewhere between ten and eleven million years ago. But the top of the hill was capped with more volcanic ash, this time forming a near horizontal layer, and dated to just over five million years old. And so this hill, insignificant to a nongeologist, was almost screaming out to me an extraordinary sequence of events. There was time enough in the geologically brief period of a few million years for the older volcanic layers and red-beds to be pushed over on end by earth movements and the resultant upheaval of the land surface to be worn away before the younger volcanic eruption covered the landscape in a new layer of ash.

The unconformities in the rocks allowed us to glimpse moments of intense activity in the mobile crust; putting all these moments in order, we began to piece together more of the story of the tilting and uplift of the rock layers during the growth of the present Andean mountains.

The Tracks in the Rocks

We can think of the emergence of the first mountains from the bottom of the vast inland sea or lake—a body of water that occupied much of what is today the high Andes in Bolivia at the end of the Cretaceous—as the birth of the Andes. Surprisingly, perhaps, there is no angular unconformity to be found marking this event. Instead, the preserved rock layers form a parallel stack, with layers of limestones, siltstones, and salt deposits at the base, laid down on the sea floor or lake bed, and an overlying thick pile of river-borne red-beds. I could use the shapes of the dunes in the earliest red-beds to locate these first mountains. However, though we could use the potassium-argon ages to put a time scale to much of their history, it has proved difficult to date the exact moment of their birth. The problem was simply that we could not find any suitable rocks to date; large volcanic eruptions were virtually nonexistent during the early history of the mountains, leaving no tell-tale potassium- rich volcanic layer behind. Thus, at first, using the few fossils that could be found in the rocks, we could only pin down the birth rather imprecisely to be somewhere in the very early part of the Cenozoic era, when the dinosaurs were already extinct after the end of the Cretaceous. Luckily for us, there was another way to date the rise of these mountains.

In 1938 two German scientists, Otto Hahn and Fritz Strassman, discovered nuclear fission—the splitting of the atom. They had unwittingly found not only the basis for an atomic bomb but a new way to date rocks. Uranium, one of the main elements in an atomic bomb, will from time to time spontaneously split. And minerals such as apatite and zircon, commonly found in granite, contain minute quantities of this uranium locked up in their crystal structures. So, when the uranium splits, fragments are hurled away with great energy, ploughing their way through the apatite or zircon and leaving behind a trail of damage. These trails or tracks were first observed when nuclear scientists started looking at the products created in their nuclear reactors at very high magnifications — fifty-thousand-fold — under the electron microscope. It turns out that it is also possible to see the tracks with an ordinary microscope if the crystal is cut open and etched with a special solution.

At high temperatures the tracks soon heal up because the surrounding atoms have enough thermal energy to reach out to their nearest neighbors and link up again. However, below a certain characteristic temperature, called the blocking temperature (about 200°C for zircon and 120°C for apatite), the tracks permanently remain as scars in the crystal. Over time, the number of these tracks will steadily increase in the crystal as more and more uranium atoms undergo fission. The rate at which this will happen can be determined very precisely from experiments in nuclear reactors. So the number of scars or tracks in an apatite or zircon crystal, recording each occasion when a uranium atom has split, can be used to calculate how long the crystal has spent at a temperature below its characteristic blocking temperature. These dates are called fission track ages.

I had come across a source of apatite and zircon on my first trip to Bolivia. This was in the Zongo Valley, which even from space shows up as a gigantic groove in the side of the Andes, and — as I described in chapter 2 — drops over four thousand meters from the edge of the barren Altiplano down into the jungle-covered foothills of the Sub-Andes. The valley cuts deep into the bedrock of the mountains, exposing large bodies of once molten granite, rich in apatite and zircon. I was interested in the Zongo Valley for one other reason. It cuts right through the region — the Cordillera Real or Royal Range — sign-posted by the ancient river dunes as the birthplace of the Bolivian Andes. Already, in the mid 1980s, Michael Benjamin and his colleagues from the U.S. Geological Survey and Dartmouth College had collected samples of the granite from along the length of the valley. Their aim was to date the granite using the fission tracks in zircon and apatite. On several occasions, I tried to extend their work by clambering among the 5,000–6,000-meter-high peaks at the southern end of the Cordillera Real, collecting more granite samples for the same method of dating. In the process, I suffered for the only time on my many trips to Bolivia the first stages of severe altitude sickness.

I had overreached myself attempting this work among the white granite pinnacles of the Three Crosses (English for the Quimsa Cruz Range) — a region that is more suitable for mountaineers. And just as we were about to retreat to lower elevations, a group of climbers, laden with crampons, ice axes, ice screws, ropes, harnesses, and carrabinas, walked into our camp. They had come all the way from Alaska and had been in Bolivia only a few weeks, acclimatizing on the relatively unknown, but lower, routes here — including the notorious Pico Penis — before moving north to the famous 6,000-meter-plus mountains of the high Cordillera Real. Once they knew that we were geologists, they cheerfully volunteered to collect rock samples from the summits. After much discussion, we finally fixed on a spot where they could hide any samples for us to pick up later. But when we did return there was no sign of either the mountaineers or the promised rocks. Perhaps they were a fantasy of my oxygen-starved brain, or perhaps you really do have to be a geologist — and a very determined one at that — to be prepared to carry heavy lumps of rock around at these altitudes?

Making sense of the fission track dates of rocks in these precipitous places requires some explanation. Imagine a vertical shaft reaching deep into the Earth, cutting through a large body of granite. Following the general increase in temperature with depth inside the Earth, the rocks become progressively hotter deeper down the shaft. Let us assume that the temperature increases at a typical rate, reaching 200°C at a depth of 10 kilometers. Rocks deeper than 10 km will be hotter than 200°C. At these deeper depths a crystal of zircon will be hotter than its fission track blocking temperature. Whenever a fission fragment, flung out from a splitting uranium atom, rips through the crystal, a scar will be created. But

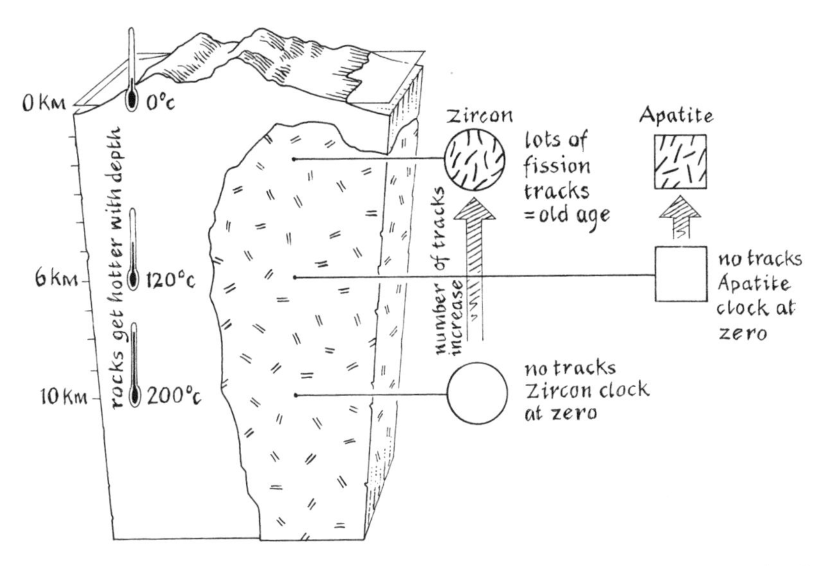

A geological clock: the splitting of uranium creates scars or tracks in crystals of apatite or zircon, but these fission tracks survive only if the crystals are below certain temperatures. By counting the number of tracks in granite, it is possible to determine the time scale of cooling in the crust—a clue to the rate at which the overlying landscape is being worn away.

these scars will heal up almost immediately, preventing their number from ever increasing; the fission track clock at these depths is permanently set to zero. What about a zircon crystal cooler than 200°C, at a depth shallower than 10 km? Here, the scars do not heal, and so the crystal becomes steadily more scarred—the fission track clock is ticking.

Perhaps the shaft was dug in a part of the world where nothing much, geologically speaking, has happened for hundreds of millions of years. There were no rivers either to remove or to add rock detritus to the surface. If one could measure the fission track ages in zircon at various depths in this shaft, one would find a very simple pattern in the ages. For all zircons shallower than 10 kilometers, there was plenty of time for the fission tracks to accumulate, giving very old ages. However, for all zircons deeper than 10 km, the age is exactly zero. On some sort of continuous plot of age down the shaft, starting at the surface, the pen of the plotter is far up the age scale until suddenly, around 10 kilometers depth, it swerves toward zero and stays there.

What happens if rivers start cutting valleys in our geologically inactive patch of the Earth, carting away rock fragments? Over time, as the rivers cut sideways and downward, the landscape is worn away, and the vertical distance down to any particular piece of rock or individual zircon crystal gets progressively less. This is a bit like an archeologist bringing the remains of a buried civilization to the surface, eventually exposing them to the sunlight by steadily stripping away the covering layers. Geologists call this process unroofing, and if it continues long enough, a once deeply buried rock is exposed in a mountainside at the surface, ready to be examined by a geologist.

If unroofing takes place very slowly, rocks at all depths in the shaft cool as they find themselves inching closer to the cold atmosphere at the Earth's surface. What about those zircons that were originally at a depth of 10 kilometers, sitting right on that critical temperature of 200°C? The moment they start to cool as they come closer to the surface, the fission track clock begins ticking. And the fission track clock in a zircon crystal at a slightly greater depth begins ticking just a little bit later, and so on. If, after unroofing over a significant period of geological time, say twenty million years, one could somehow measure the ages down the shaft, one would find that the pattern had changed from before. On the continuous age plot, the ages now swerve from high ages to an age of twenty million years, then decrease more gradually to zero at a depth of 10 kilometers: the swerve in the plot defines the moment when the rivers started eroding the landscape. But there is more. The slope of the plot, thereafter, is a measure of the rate of unroofing or how fast the landscape is being worn away.

The behavior of apatite in our imaginary vertical shaft is virtually the same as that of zircon. The only difference is that the zero-age depth now lies at about 6 kilometers, where the temperature is 120°C. Taken together, the two minerals give geologists a more complete picture of the unroofing history. So how does all this apply to the zircon and apatites from the Cordillera Real, in the Zongo Valley and Quimsa Cruz mountains? The huge topographic relief in the region between the highest and lowest outcrops of granite, exposed in such a short horizontal distance from the mountain peaks to the lowest levels of the Zongo Valley, has bared deep levels in the crust. It is as though one wall of our vertical shaft has been prized away, so that we have a view of the other wall, extending down nearly 4 kilometers. As we shall see, the fission track ages of the apatites and zircons at various levels in this wall of rock reveal a history of the mountains since they first started to form.

Watching the Mountains Grow as They Are Worn Away

The birth of the earliest mountains in the Bolivian Andes set in motion forces that would start to wear them away. The new mountains attracted rain. The rain cascaded off the mountainsides, wearing away the bedrock and carrying the rock fragments to the lowlands. This erosion triggered the fission track clock, and we can read the clock by looking at the fission track ages from the rock in the sides of the Zongo shaft. Michael Benjamin's fission track dates for zircons, at that critical swerve in the age plot down the shaft, show that this began to happen around forty million years ago, in the Eocene period. Thereafter, as the fission track clock ticked, more rock was removed from the mountains, and the pile of red-beds in the lowlands became steadily thicker. So, curiously enough, we can watch the mountains grow only by looking at them being worn away.

The zircon and apatite fission track ages show that on average, since roughly forty million years ago, the mountains have been worn down at an average rate of a third of a millimeter each year. This may not mean much unless one considers that tens of millions of years, by human standards, are an extraordinarily long period of time, long enough, in fact, for erosion at this rate to remove all the rocks extending down to a depth of a staggering 15 kilometers.

Today, this region is, on average, about 4,000 meters high. So how could it have reached this height if a 15-kilometer-thick slab of rock has actually been taken away? The answer lies in the forces that push up mountains in the first place, counteracting the power of rivers to wear them away. The actual rise of the mountains is a battle between these two agents of change—a battle that can profoundly affect the evolution of a mountain range. If the two are in balance, the mountains stay at more or less the same height, attaining a sort of steady state. But if one is greater than the other, the mountains may either rise or fall.

We are now in a position to watch, as it were, the evolution of the Andes in Bolivia and northern Chile since the end of the Cretaceous. The various lines of evidence that I have talked about in this chapter—the shapes of dunes in the river-borne red-beds, potassium-argon ages of volcanic layers, unconformities in the rock layers, fission track ages in the granite bedrock—when put together, allow us to trace the evolution of the mountains from their birth, giving us frozen frames in the film of their continuous growth as they become both higher and wider. From a human perspective this has happened almost imperceptibly, though the individual earthquakes, which are the heartbeat of growth, have created their own brief moments of terrifying havoc. Geologically speaking, however, the mountains have risen rapidly. To visualize this growth, imagine being a space traveler who keeps returning to the lowland plains of what is today the Amazon jungle—a region that so far has remained unaffected by the vast upheavals farther west. The time scales involved, spanning tens of millions of years, may seem both daunting and difficult to comprehend. They may, however, have more meaning if placed in the context of the evolution of our own species along its branch of the tree of life.

On a first visit, just before the end of the Cretaceous and slightly more than sixty-five million years ago, you see a vast inland sea or lake on the western margin of South America. Looking westward, across this body of water, you can make out isolated volcanoes on the far shore in the distance, and beyond these the Pacific Ocean. You even spot a herd of dinosaurs moving slowly along the shoreline and watch a plume of ash spew out of an erupting volcano. About forty million years ago, the first important branch in our evolution had sprouted with the appearance of the nocturnal tree-dwelling primates. But the dinosaurs are long extinct and never saw the Andes. At this time, on a second visit, you find that the lake water has gone. Instead, you notice a long, low ridge rising above the old lake bed, with a lowland plain, either side, traversed by wide rivers. Thunderstorms occasionally hang over the ridge, rumbling and flashing with lightning and releasing heavy rain. The rain runs down the ridge flanks, feeding the rivers on the plains.

By the time of your third visit, thirty million years ago, the first monkeys have begun leaping from tree to tree during the day. Now, the ridge is gouged by rivers, forming a range with gullies and valleys. River detritus, accumulating at the eastern front of the range, has built up vast fans of sediment. Another system of rivers, flowing westward off the back of the range, has dumped detritus into a more distant wide lowland plain—this is the region that makes up the high Altiplano today. Even farther west, those once isolated volcanoes are a continuous chain of volcanic peaks, building a second long, narrow range of mountains that acts as a barrier between this lowland plain and the Pacific Ocean.

On successive later visits, you witness the two mountain ranges becoming ever more prominent. River detritus from the rising mountains continues to pour into the intervening plain, raising the land surface as layer upon layer of red-beds accumulate. About twenty million years ago, soon after the appearance of the great apes—the first hominoid primates—the eastern mountains appear closer to the heart of South America, having advanced about a hundred kilometers or so farther eastward during the previous five to ten million years like a gigantic wave of rock. Old river beds, which once traversed the lowland plains, are caught up in the mountain-building, lifted hundreds of meters, and tilted and faulted, preserved now as only small fragments of what they once were. At about ten million years, you find that the two early mountain ranges are touching each other, coalescing to form the high Altiplano. This is not long before the human lineage, beginning with the hominids,

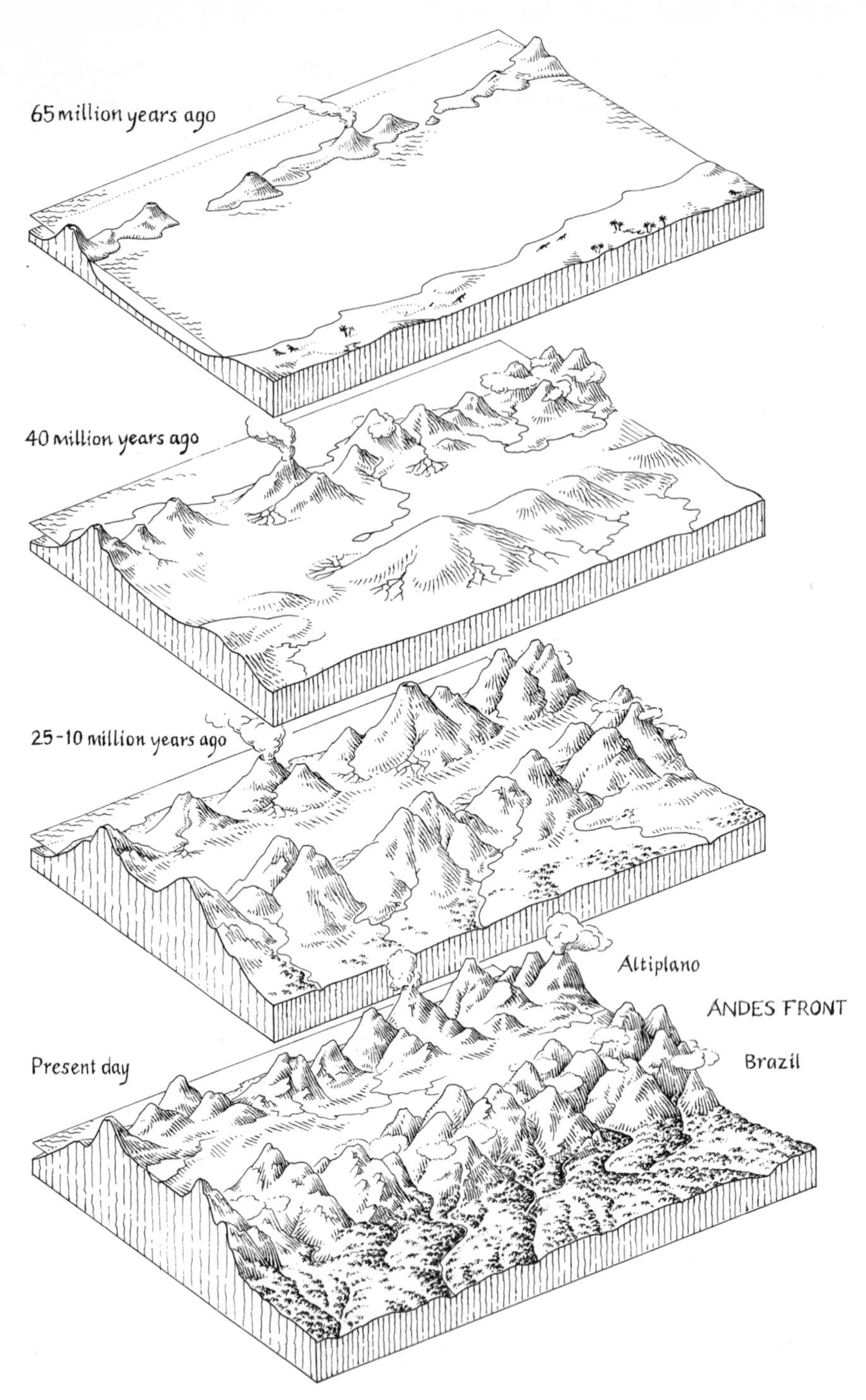

The story in the rock layers: a series of snapshots of the Bolivian Andes, viewed from the lowland plans in the east, showing how they have risen during the last sixty-five million years.

branched off from the apes. Since then, as hominids in Africa first started to walk upright on two legs, eventually evolving into modern humans, the mountains have risen yet higher, advancing a further hundred kilometers or more eastward into the lowland Amazon jungle.

And so we come to the end of our journey through time. Using not much more than a hammer and compass, it has been possible to peer into the past and see snapshots of these mountains at various stages in their growth. Dating these snapshots has proved a little more difficult, and we have had to exploit the atomic nature of rocks. These images show what has happened at the surface, covering a time span that turns out to be similar to our own evolution from the earliest primates. In fact, it may not be pure chance that the two time scales are so similar—something I will touch on in a later chapter. However, there is much more to mountain building than can be observed above ground. In the next chapter I will explore the evidence for what has happened much deeper down, in the roots of the mountains. First, though, as a sort of sideshow to the main scientific story, let me tell you some more of the day-to-day realities of my own field work in Bolivia; I will carry on with this theme at the end of subsequent chapters.

⁂ ⁂ ⁂ ⁂ ⁂

Campesinos and Blockades

The people of Bolivia come from several ethnic backgrounds. In the mountainous Andean regions, the indigenous peoples make up the bulk of the rural population—referred to collectively as *campesinos* (literally, country folk)—who are either Aymara or Quechua speaking; both are the direct descendants of the Incas. Most campesinos speak Spanish as a second language, though it is still not unusual to come across women in remote villages who do not understand Spanish. In the tropical lowlands, there are numerous isolated tribes of Amazonian Indians, each tribe having its own language. Finally, in the towns, most Bolivians are of mixed Andean and European (mainly Spanish) origin. The main political division among all these peoples is really one of wealth, between the relatively rich city dwellers, who have largely adopted a modern Western lifestyle, and the extremely poor rural population, who, to a large extent, live in a way that has remained unchanged for hundreds of years. The ever present tension between the city and country people lies at the heart of almost all Bolivian political issues.

There seems to be a strong sense of cultural identity among the campesinos, despite the isolation of many rural settlements, and they are surprisingly well organized. Some of this stems from the socialist element in twentieth-century Bolivian politics, a direct consequence of the large mining community. Many trades are highly unionized, and the locals are capable of acting as a unit. The campesinos have a number of particular grievances. They see the bulk of the wealth in Bolivia going to the cities with little benefit to themselves. They have a difficult and uncomfortable life and have to put up with poor schools and hospitals. And they feel that their culture is being threatened. In particular, the American military campaign to wipe out the coca plant—the raw material of cocaine—is stopping an age-old tradition of growing and consuming coca leaves. Finally, they have suffered a considerable rise in the cost of transport because of recent fuel price hikes.

All this has made the campesinos particularly volatile and liable to take direct action at any time to bring attention to their grievances. The commonest form of this action is a blockade of the road system. While we have been working in Bolivia, during the last ten years, there have been a number of these blockades. They usually involve blocking either the main road between major towns or all of the entrances to a particular town. One time, we arrived in the Altiplano town of Uyuni, after several weeks in the field, just before one of these blockades was set up. The local population in the entire southern province of Potosi had called an indefinite general strike, and all road movement was going to be stopped. We were not thrilled by the prospect of being stuck in Uyuni for several weeks. It turned out that most of the inhabitants of Uyuni were also unhappy about it—it would ruin local business—and there were many people who were ready to run the blockade. While we were waiting for a tire puncture to be fixed, an old man, who wanted to go home himself, offered to show us a way out of town.

Uyuni is a small town of less than a thousand people. It is built on a grid plan at the junction between two important rail lines. The main roads out of Uyuni radiate in four directions. Our proposed route was merely a side street that left the town at a "corner" between two of these main roads. We could clearly see the blockades as we left the town limits. The people manning them soon saw us and started running in our direction. We had a Bolivian geologist with us on this trip, so I asked him what we should do now. If the strikers reached us it could be very dangerous—campesinos are not adverse to stoning anybody who seems to be defying them. The Bolivian geologist suggested that I go faster, which I did. We left Uyuni in a hurry, bouncing along the rough track and kicking up a large cloud of dust. We soon outpaced the strikers and did not stop until we had crossed the border with the neighboring province of Oruro, which was not on strike. It was only then that we fully appreciated the risk that we had run.

Another time we were trapped in Sucre, the old capital of Bolivia in the Eastern Cordillera, during a nationwide blockade of the road system. The only way in or out of the major towns was by air; the Bolivian Army was keeping the lifelines between town and airport open. Each day we would venture out to the police barrier at the city limits to get the latest news. Nobody really knew what was happening, and we would get conflicting reports: the road was open, the road was closed. Once, we did try to leave Sucre by road. About ten kilometers out of town we came across large boulders on the road. These became more frequent, and there were groups of campesinos gathered on the roadside. They seemed particularly menacing to us because we had just heard that a driver had been stoned to death when he had tried to break through a blockade. We nervously turned around, hoping that this would not precipitate an attack, and retreated to Sucre.

By a strange coincidence, while all this was going on, I was reading a history of the battle for Stalingrad during the Second World War, when the German Sixth Army was encircled by the Russians and trapped in the town. During our confinement in Sucre, as the days went by, I began to have nightmares in which Stalingrad and Bolivia somehow got mixed up. This state of mind was not helped by news reports claiming that the campesinos were moving into the suburbs of La Paz, tightening the noose around the similarly blockaded city. Eventually, we managed to get on a flight out of Sucre. But we were going to La Paz; out of the frying pan and into the fire, I thought. As we made our approach for landing, the plane swooped low over the blockaded main road into La Paz, strewn with rocks and cut by huge ditches. It would have been an easy matter for a group of overzealous campesinos to damage the aircraft by throwing rocks at it with hand slings. When we came to leave La Paz, we had to get to the airport in the middle of the night because it was rumored that an "army" of campesinos was planning to cut off the airport road early in the morning. When the plane finally climbed high into the air and headed north to Miami, my feeling of relief was enormous.

✣ CHAPTER SIX ✣

Putting Down Roots

Root, ***n.*** Part of thing attaching it to a greater or fundamental whole. *(OED)*

THERE IS A RESTAURANT in the back streets of Vienna that specializes in schnitzel. I remember going there one year during the Autumn Heurigen festival when the first wine of the year is ready for drinking. The tables were overflowing with noisy groups of people who had come to enjoy the new wine and schnitzels. I soon realized what was special about the restaurant when a waiter rushed by with several plates balanced on his arm; gigantic slabs of veal, coated in breadcrumbs, were literally hanging over the edges of the plates and flapping like loose cloths. My companion, Leonore Hoke, told me that these were the largest schnitzels in Vienna. Yet they were surprisingly cheap. The chef must have gone to unusual lengths to flatten the lumps of veal, pressing them into thin slabs or sheets—I could imagine him behind the kitchen door pounding the meat with a wooden mallet. Curiously enough, these thoughts about schnitzel flattening triggered memories of my experiences earlier that summer, during my field work in Bolivia.

I had been searching for outcrops of limestone in the Eastern Cordillera, following up some information from a Bolivian geologist, who had told me that there were limestones near the village of Otavi, not far from the main road south. When we reached Otavi—I was traveling with Leonore—we found that it lay on a dusty apron of land at the foot of towering sandstone cliffs, which seemed to form a gigantic wall in the landscape. A small track led on from the village toward this wall, and apparently no farther. It turned out that the trail passed through a narrow, sinuous ravine, following the course of a stream, and the limestones were farther upstream. We slowly negotiated our Land Cruiser along its bouldery bed. Around a corner, we suddenly emerged from the ravine into a hidden world—a long, narrow plain, about a kilometer across, surrounded in all directions by cliffs of the same sandstone. We set up camp in the middle of this plain, on a cushion of mossy grass near the stream, and later that afternoon, I wandered off to see if I could find the limestones.

Everywhere I looked, I found signs that the once horizontal rock layers had been tilted up on end. As I tried to visualize this, sketching ideas in my notebook, I realized that I was standing in the middle of what was, in effect, a gigantic canoe or boat, a few tens of kilometers long, made of great twisted slabs of rock. Vertical sheets of the resistant sandstone and limestone formed the hull of this canoe, rising up all around and walling off this strange place from the rest of the Eastern Cordillera. And the softer layers of siltstone and shale, which once filled the middle, had been worn away by the stream, leaving behind the hollow, boat-shaped shell. It was as though the rocks had been caught up in a vice, pressed and molded by vast forces so that the once horizontal layers crumpled up, twisted right around, and bent into this new form. Intense squeezing or squashing like this inside the Earth had become linked in my mind with the schnitzels.

This takes us right back to the fundamental problem of why and how mountains are created in the first place—the problem I had wanted to solve from the very beginning of our Andean project. In an earlier chapter, I told the story of the peculiar behavior of a surveyor's plumb line during a survey in the 1850s of northern India, immediately south of the Himalayas. George Airy and John Pratt—the former a surveyor and the latter a mathematician and clergyman—had both tried to explain this by speculating on what lay beneath the high peaks of the Himalayas. The acceptance of one or another of their explanations had led to widely different theories about the origin of mountains. Airy and Pratt both believed that the outer part of the Earth formed a crust that rested on a deeper interior called the mantle. Airy had proposed that the crust beneath the mountains was like an iceberg with deep roots—where the mountains were higher, the crust was thicker, with roots that reached down deeper into the underlying mantle. In marked contrast, Pratt had proposed that it was not the thickness of the crust that changed much, rather it was the crust itself: it was less dense beneath the mountains, like risen bread, compared to more dense heavy dough in the lowlands. By the late nineteenth century, geologists were already drawn up in opposition, along battle lines inspired by these two very different ideas, though, in fact, neither had been conclusively proved for any part of the Earth.

One might have thought that over a hundred years later the battle would have been long fought and won. After all, the modern theory of plate tectonics had swept through geology, blowing away the dusty cobwebs of the past. And to a large extent, in the 1960s and 1970s, the nature of the Earth's crust was well known. Beneath the sea floor it was quite thin with a remarkably uniform thickness, like the skin of an onion. Pratt was basically right when it came to explaining the variation in the depth of the oceans in terms of changes in density. On land, beneath the continents, the crust was certainly much thicker, rising high above the sea floor. Here, it seemed that Airy's idea of crustal thickness was the one that made more sense. But a precise image of the crust beneath high mountainous parts of the continents seemed to be more elusive. It was proving difficult to access these regions, and the sort of research required to probe deeply into them was likely to be very expensive, consuming millions of dollars. Geologists had to be content with what they could see on the surface. The observations remained conflicting.

The Andes were particularly confusing to a geologist. There was no doubt that the rock layers had been contorted here—they were tilted and broken, like the layers in my rock canoe. Thus it seemed likely that there had been squeezing of the crust, and perhaps this was enough to thicken it and push up the iceberglike mountains that Airy believed in. But there was also abundant evidence for volcanic activity, possibly a sign of significant heating and expansion that has reduced the rock density beneath the mountains in just the way that Pratt had envisaged. The products of this volcanic activity—

intrusions of granite and other once molten rock—could also have added to and thickened the crust, thereby raising mountains more in line with Airy's ideas, but without the need for squeezing. Certainly, impressive masses of volcanic rock and granite were clearly visible, well known, and even burrowed into—they were the source of some of the world's largest deposits of copper and tin and so could hardly be ignored. Indeed, at the beginning of the twentieth century, European miners knew from bitter experience about the Andes. The Bolivian mines had forced down the prices of many important metals and put them out of business, making the Bolivian tin baron, Simon Patino, one of the richest men in the world. In the centuries before this, the almost unlimited supply of silver from Bolivia's Cerro Rico—an ancient volcano—not only financed the Spanish Empire, but underwrote the cost of numerous expensive seventeenth- and eighteenth-century European conflicts.

The outcome of all this was that when I first started working in the Andes, there was a general impression in wider geological circles, depending on who you spoke to, that the Andes were either mainly volcanic and had been raised up by volcanic activity, or had been squeezed up where two great tectonic plates were relentlessly moving toward each other. The obvious way to assess these two views of mountain building was actually to measure the thickness and density of the crust beneath the Andes. There was no chance, however, of doing this directly. Despite the best efforts of oil companies, the deepest drill holes only scratch the surface, penetrating no more than a few kilometers. Geologists were going to have to be much more cunning and ingenious. In this chapter, I tell the story of how our Andean project, building on the work of many others, has tried to probe deep beneath the Bolivian Andes. At times this may sound a bit like a long shaggy dog tale, with its many twists and turns. But, if you bear with me, you will find out how vast mountainous tracts of the Bolivian Andes have been raised to more than the height of the highest mountains in the European Alps or Rocky Mountains, in a mere blink of the long geological history of South America. The story begins, not in the Andes, but in Renaissance Italy.

Falling Objects

Around 1600 Galileo Galilei invented an ingenious device to study how objects fall to Earth. I once watched a demonstration of this, though I am not sure if what I saw was identical to Galileo's original experiments. The person giving the demonstration had carved a deep, straight groove in a long plank of wood. The plank was then tilted to form a ramp, so that the groove sloped downhill, and little bells were suspended above it at regular intervals along its length. A metal ball, released at the top end of the slope, would roll down the groove, hitting the bells and causing them to ring. It was quite noticeable that the individual chimes got closer and closer together as the ball passed successive bells. It was obvious that the ball was traveling faster and faster down the slope, accelerating as it went.

Galileo, in his original experiments, had also come to this conclusion. Nearly a hundred years later, Isaac Newton explained this acceleration with his theory of universal gravitation. He proposed that any object would fall to Earth, accelerating as it went, because it was being pulled by the mutual gravitational attraction between the mass of the Earth and the mass of the object. However, an important feature of Newton's theory is that the actual acceleration caused by this attraction does not depend on the mass of the free-falling object. This prediction, which comes as a surprise to many people, was already credited to Galileo, who is said to have simultaneously released a cannon ball and musket shot from the top of the Leaning Tower of Pisa. The two balls are supposed to have hit the ground at virtually the same time, much to the amazement of many of Galileo's fellow scientists.

Newton showed that it was possible to calculate fairly accurately the acceleration of a free-falling body if one knew both the total mass of the Earth and the distance of the body from the center of the Earth—the greater this distance, the less the acceleration. The weightlessness of astronauts in space is graphic demonstration of this reduction in acceleration far away from the Earth. But this is not all. The rate of acceleration at any point on Earth is also affected, though to a much lesser extent, by both the shape and spin of the Earth, and the distribution of massive rocks deep inside. So, very accurate calculations would need to take all these factors into account. However, the important conclusion remains that the acceleration of a free-falling object is a fundamental property of the Earth and not the falling object. Physicists often refer to this acceleration due to gravity by the letter g. At the Earth's surface, it is roughly the same as accelerating from 0 to 96 kilometers per hour (60 miles per hour) in slightly less than three seconds—you would need a very powerful car to be able to accelerate like this.

It was during a famous French survey of part of the Ecuadorian Andes, in the late 1730s, that a junior expedition member, Pierre Bouguer, saw a way to use *g* to probe the Earth. The survey had been trying to measure the shape of the Earth with accurate leveling instruments. But they had completely ignored the gravitational pull of the massive rocks inside the mountains. Bouguer realized that this could disturb the sensitive leveling instruments, causing errors. Measuring *g* would be a way both to correct these errors and to work out the mass of rocks in the mountains. If this mass is tightly concentrated, the rock has a high density, but if the mass is more spread out, the density will be lower. And these densities depend, to a large extent, on the composition of the rocks, so the local value of *g* also reflects the type of rock—for instance, the presence of rocks from either the crust or the mantle. Thus we reach the thought-provoking conclusion that just by dropping a ball on the surface of the Earth, we can actually begin to understand the nature of the rocks deep inside.

A Red Herring

If you want to buy a detailed topographic map of Bolivia, don't expect to be able to get one in a shop. You will have to brave the main headquarters of the Bolivian Army in La Paz—the Estado Mayor. The first time I went there, I had an appointment, arranged through the Jesuit director of the Bolivian Seismological Observatory—the

Observatorio San Calixto — to see the colonel in charge of mapping operations. I was escorted across the base by two armed soldiers, who handed me over to another soldier at the entrance to a rather grim-looking building sprouting antennas and aerials from every conceivable part of the roof. Up some stairs, I passed through double doors to enter what seemed to me like another world. There was a hum of air conditioners, and in front of me, the American flag, hanging from a pole, was reflected in the highly polished floor. I got my map, but somehow, in the process, I had wandered into U.S. territory, right in the heart of a Bolivian Army base. The reason for this soon became clear.

Not long after the Second World War, South America began to assume a strategic importance to the United States. It became a military priority to obtain detailed information about the nature of the terrain. And so in the early 1950s the United States Army sent teams of surveyors out to almost all the South American countries, including Bolivia, to start the huge task of making topographic maps. The maps that were subsequently produced — especially those in Bolivia — are a major achievement. The heavy survey equipment had been carried up to the tops of many of the high peaks, often in difficult weather conditions, and survey lines had been hacked through the thick jungle of the lowlands. As part of this undertaking, detailed measurements of the acceleration due to gravity — g — were also made; these would help define the base level for the maps. And it is these measurements, especially when combined with the information from the topographic maps, that have helped us look beneath the Andes.

Whenever geophysicists want to observe gravity, they use an instrument called a gravimeter. They might set it up on a wide, flat plain or in a deep valley, or on top of a mountain. The actual observation is a very fussy affair, but it is not necessary to drop an object. Gravimeters are ingenious and delicate instruments full of carefully balanced weights and springs that do the same job, but they are so sensitive that they can easily detect a difference in g between your kitchen floor and kitchen table, a difference that exists only because the table top is farther than the floor from the center of the Earth!

It is only tiny variations in g that reveal the nature of the bedrock. For these to become clear, the easily calculable effects of the shape and spin of the Earth and height of the observation point, as well as one or two other minor factors, must be stripped away from the gravity observation. The final observation, calculated this way, can then be compared with a prediction based on the assumption that the Earth is just a smooth ball of rock. Geophysicists have had many international conferences to agree on what this predicted value should be. Any difference between an observation and the prediction is called an anomaly (in fact, the technical term here is a "free air anomaly"). Over the years, the original work of surveyors in South America has been added to by industry and academic geophysicists, revealing a remarkable result. The free air anomaly in valleys and on mountaintops right across the Bolivian Andes is nearly everywhere surprisingly small, especially in the heart of the ranges where the mountains are highest—despite their vast mass, the mountains do not seem to have much effect at all on g. So what are we to conclude from all this? Well, as I shall explain, it is proof that one or other of the models for the crust proposed by George Airy and John Pratt nearly 150 years ago must indeed be correct for the Andes. The question is, which one?

Both Pratt and Airy were making use of the ideas of the Greek philosopher Archimedes, worked out over two thousand years ago. Archimedes, when he leapt out of his bath and exclaimed "eureka" (so legend has it), had cracked the problem of how and why a ship (or any other object) sinks or floats. In Pratt's and Airy's day, many scientists thought that the deep and hot interior of the Earth—what we now call the mantle—was a liquid (more accurately a fluid). In this case, the overlying solid crust of the Earth must be floating on this fluid, finding its own level like a ship at sea. Using Archimedes' principle, both Pratt and Airy could rely on a simple idea: if we could somehow extract and weigh a series of rock columns extending down through the crust to some constant depth in the underlying fluidlike mantle, we would always find that they had the same weight. This will remain true, regardless of what part of the mountain range, or adjacent lowlands, the rock columns come from. In Pratt's view, because he thought the base of the crust was at a constant level, it would only be necessary for the rock columns to reach down this far: as the mountains got higher, so the density of the crust lessened, thereby keeping the weight of the columns the same. But, according to Airy, the density of the crust was uniform, but the thickness varied. In this case, his rock columns would have to extend much deeper down than Pratt's, to a constant level in the mantle below the roots of the thickest crust.

It turns out that it is the equal weight of neighboring columns of rock, in both Airy's and Pratt's view of the crust, that will always ensure that the free air anomaly is small, regardless of the height of the mountains. In other words, the gravity measurements across the Andes are consistent with both Airy's and Pratt's models, but have failed to distinguish between them. At this point, you may feel that all the gravity work has been a bit of a red herring—but, at least you know now the reason for the heading I have given to these paragraphs! It may seem that we have not advanced very much in our understanding of the Andes. But, believe me, we have. Despite the fact that geologists had enthusiastically latched on to Airy and Pratt, we are still dealing just with ideas, and these ideas had been originally developed to explain some features of a survey near the Himalayas. The gravity measurements have now elevated these ideas to the status of front-running models for the crust beneath the Andes. In fact, we have only just begun to make use of g in the Andes. Later in this chapter I will show you how gravity enabled us to work out in great detail the mechanism by which large tracts of the Bolivian Andes were raised. But this would not have been possible if there had not been other ways to probe the crust, ways that would prove crucial to solving the mystery of its nature beneath the mountains. We still needed to decide between Airy and Pratt, and this distinction, as I alluded to at the beginning of this chapter, had profound implications.

Were we dealing with a mountain range that had been built up by volcanic activity—perhaps dragged up by the rise of vast volumes of low-density and buoyant granite and other intrusions of once molten igneous rocks—an idea effectively implied by Pratt's model?

Or had it been pushed up as the crust was thickened by horizontal squeezing, as suggested by Airy's idea? Or, perhaps, it really was some combination of the two. It was not until several years after I had started working in Bolivia that I was in a position to properly answer these questions.

Good Vibrations

In 1995 I was invited to Cornell University in New York as part of a panel to assess various proposals to examine the crust in the Andes. I was venturing into the very heart of the large Cornell Andean project—I had the distinct impression that some of its members saw me as competition. The Cornell group, lead by Rick Allmendinger and Bryan Isacks, had a long and distinguished track record of geological research in the Andes of northern Argentina, though with some sorties into Bolivia, fielding numerous students. In their eyes, I think, our much smaller and more youthful project had not yet earned its stripes. They had a number of seminal scientific papers under their belts. And like all good researchers, they had plenty of ideas. I hoped that our small project would soon be in a position to make an important contribution as well. We were rapidly accumulating a surprisingly large cache of geological facts. I also had the luxury of being able to spend long periods in the field, using some of this time to visit areas that had been the subject of previous studies, checking field relations and trying to convince myself that the published conclusions made sense. I found myself reluctant to accept any idea until I had seen, at first hand, the evidence for it.

The Cornell conference forced the various South American groups to rub shoulders and, at least outwardly, to appear friendly, though I sometimes sensed an air of caution when the conversation got around to hard data. However, the panel members had agreed to give talks on their work. The talk immediately before my own was given by Susan Beck, an American seismologist from the University of Arizona. Susan radiated efficiency and precision. As her talk unfolded, I began to realize that she had worked out in great detail the thickness of the crust beneath the Andes. My own talk seemed to pale into insignificance. She described how in the previous year she had been part of a group that had laid out a line of sensitive instruments for listening to earthquakes—seismometers—across the Bolivian Andes. By chance, the group was able to record an unusually deep earthquake that occurred beneath the lowlands of Bolivia at a depth of nearly 700 kilometers. Vibrations, spreading out from this earthquake like circular ripples in a pond, traveled upward through the mantle and crust toward the surface. It is well known that the first waves to arrive at the seismometer are essentially sound waves, vibrating back and forth like a spring—these are called P waves (P for primary). Later, another type of wave, with a more undulating movement, reaches the seismometer, having traveled through the Earth slower than the P waves. These are the S waves (S for secondary).

The earthquake had provided Susan Beck's group with a shortcut to working out the thickness of the crust. Rather than having to go through the involved and expensive procedure of letting off their own explosions, they could let the Earth do the work for them and just sit back and listen to the results. In fact, they were listening to the outcome of a remarkable journey through the Earth. The P wave seismic energy travels upward from the source of the deep earthquake until it reaches the Moho. Here, at the boundary between the mantle and crust, there is a change in both the composition and properties of rocks, and so the P wave slows down. The reduction in speed has a curious effect, causing some of the P wave energy to be converted into an S wave vibration. The remaining energy continues on as a P wave, eventually reaching a seismometer at the surface. A short while later, the converted S wave, which has traveled more slowly through the crust, also arrives. Susan spotted these short time delays, generated by the P to S wave conversions at the Moho, in the characteristic pattern of squiggles that make up the seismic record of her seismometers. The journey of these two waves is a bit like a race between two athletes. The P wave is always faster than the S wave, and so its overall lead at the end of the race will depend on the length of the course—the longer the course, the more time the P wave will have to establish its lead. The length of the course is merely the thickness of the crust.

Each of Susan's instruments recorded its own seismic time delay, in effect registering the results of the local seismic race in the underlying crust. It was immediately obvious that the time delay was greater in the interior of the Andes than at the edges. Using the known speeds of S and P waves in the crust, it was possible to convert this time delay directly into a crustal thickness. From this, Susan's team found that the crust did, indeed, thicken toward the high parts of the Andes, and the base of the crust at the Moho was almost a perfect exaggerated reflection of the general shape of the mountains themselves. The thickness of the crust nearly doubles from about thirty-seven kilometers beneath the lowland plains to over seventy kilometers beneath the high spine of the Eastern Cordillera, where individual peaks rise up over 6,000 meters above sea level.

I realized after listening to Susan Beck's talk that the towering peaks of the snow-capped high Andes really are just the tips of a rocky iceberg, just as George Airy had predicted. There were indications of this in previous attempts to detect the base of the crust, relying on the phenomenon of refraction or bending of seismic energy, but the results had been disappointingly vague and imprecise. Susan Beck's clever method had now brought the base of the crust—for at least one part of the Bolivian Andes—sharply into focus. Beneath the mountains, deep roots in the crust reach down into the Earth's mantle, and for every kilometer that the peaks rise up, the roots extend about seven kilometers deeper. But Airy's rival, John Pratt, was not completely knocked out, for, in reality, we would not expect the crust to be as entirely uniform as Airy originally imagined. There will be inevitable variations in composition where bodies of granite or basalt have intruded, or where there are thick accumulations of sedimentary rocks. These compositional differences will certainly result in local changes in density.

One positive outcome of the Cornell meeting was that a team of geophysicists from the Free University of Berlin, supported by the apparently inexhaustible resources of the German Science Funding

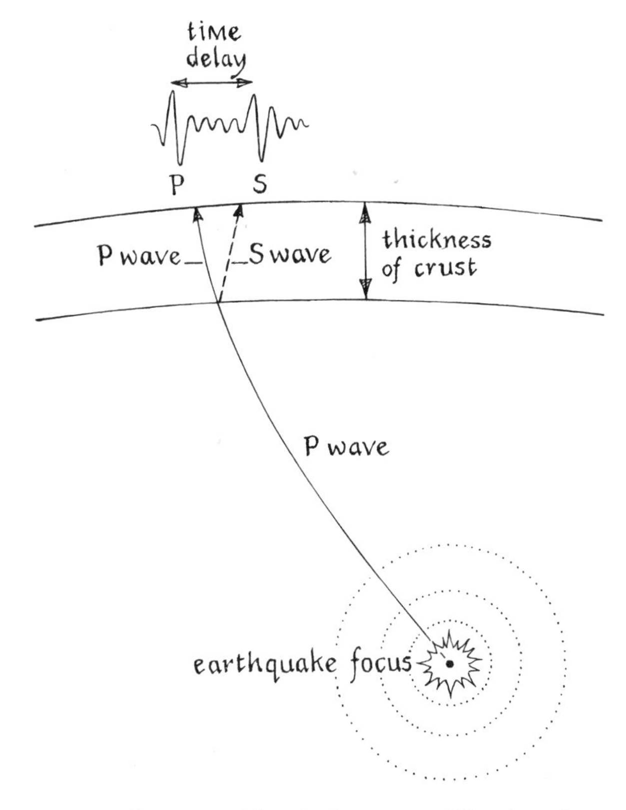

Probing the crust with seismic waves. Vibrations from a deep earthquake turn into two distinct types of wave when they pass through the Earth's crust. The S wave is slower than the P wave, so it reaches the surface slightly later: this time delay can be used to calculate the thickness of the crust.

Agency, finally decided to probe the crust right across the Andes, in northern Chile and southern Bolivia, with the latest techniques developed by the oil industry. Seismic vibrations from man-made explosions were picked up by many seismometers. In this way, they managed to determine differences in the speed of the seismic waves as they travel through the rocks—a sound scan of the crust. Some years later, their project leader, Peter Wigger, told me that they had risked a diplomatic incident when they set off explosions and laid kilometers of cables across the international border between Chile and Bolivia—two countries that were virtually at war with each other. I saw the results of all this at another conference—this time in Chile—in 1997. The German group had also managed to work out in detail the density structure of the crust beneath the Bolivian Andes, in effect mapping out the different types of rocks. They could do this because there is a good correlation between the density of rock and the speed at which the seismic vibrations travel through it. And as a final trick, they could bring the image much more sharply into focus by refining the density estimates so that the crust pulls its weight, so to speak, in agreement with the existing measurements of gravitational *g* across the Andes. Airy's icebergs are still there, but they are not, after all, pure "ice"; they contain other impurities.

After Cornell, I flew on to Bolivia for some more research of my own. Overall, I was grateful for the past few days. I had nearly declined the offer to go to Cornell—my field work in the Andes seemed a bigger priority. I knew now that what I had just learned at the conference was going to be very important for my research.

Smoothing Out the Creases

My musings on Viennese schnitzels at the beginning of this chapter graphically illustrate a way to make the mountains of the Andes: if the volume of crust is to be preserved, then squeezing or squashing it one way will cause a corresponding thickening or expansion in another direction. Squeezing the layer of the Earth's crust, therefore, in a horizontal direction should thicken it vertically, pushing up mountains and pushing down roots; it is a simple matter to calculate the amount of squeezing that would be needed. And so one way to test this idea is to measure the actual amount of squeezing in the crust beneath the mountains and see whether it is enough to make the mountains and their roots.

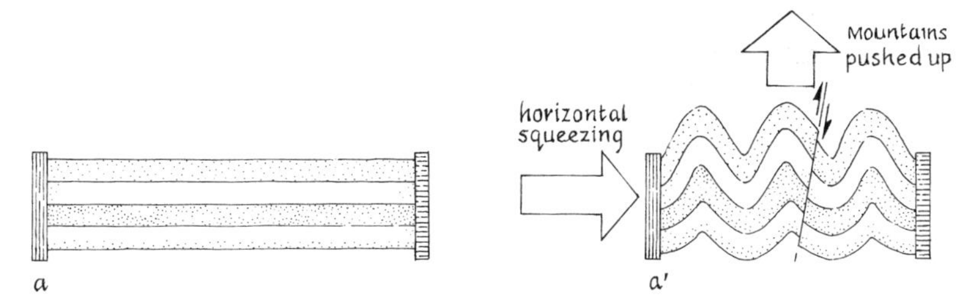

Smoothing out the creases. Horizontal layers of rock are contorted when the Earth's crust is squeezed. The amount of squeezing can be calculated by measuring the length of the folded and faulted layers.

Working out the actual squeezing in the crust beneath a mountain range is rather like attempting to smooth out a heavily crumpled or creased sheet of paper. In the case of the Andes, the crumpled piece of paper is analogous to a layer of rock strata. When this layer was originally laid down, it would have been like a sheet of flat paper. Squeezing of the crust in one direction has buckled and broken this layer, creating a series of parallel folds and faults. By carefully measuring the length of the strata, around every bend and buckle, and linking up broken layers across faults, it is theoretically possible to estimate the original horizontal dimensions of the piece of rock "paper." However, this method relies on knowing every buckle and fault in the rock layer—missing some of these could seriously affect the calculation. In addition, one has to find the right sheet of rock, one that had been laid down flat just before the mountains started to rise.

On my first visit to Bolivia in 1989, Ramon Cabre—the director of the Observatorio San Calixto—told me that a young American geologist, Barbara Sheffels, from the Massachusetts Institute of Technology in Boston was already making a new map of the Bolivian Andes using photographs of the mountains. I remember feeling alarmed: would our project be duplicating work that had already been done? Ramon had clearly provided her with an enormous amount of help. He sighed with relief as he described how he had sent off the last consignment of photographs to Sheffels. These had been extracted from the Bolivian military with great difficulty. They had been taken from the air and clearly revealed the rocks on the mountainsides of the arid Eastern Cordillera, laid bare by the action of rivers and glaciers. Sheffels was using photographs taken from slightly different viewpoints to see the landscape in three dimensions. In effect, looking at two neighboring photographs is like viewing the mountains themselves with two eyes, allowing the brain to form a stereo image, so that lines in the rocks on individual photographs stand out in relief, revealing the full shape of the strata. Barbara Sheffels had examined hundreds of photographs, working out the buckles and faults in the rocks. In this way, she filled in the gaps in the existing geological maps and then tried to estimate the original width of the once flat sheets of rock.

When I finally came to see the results of Sheffels's work I realized that she had faced many problems. The geological history of the Bolivian Andes was still only poorly known, so it was often not clear which rock layers to study. In any case, there was no easily identifiable layer that could be traced right across the mountains, and so there were always going to be gaps. Where the recognizable layers existed, it was sometimes difficult to know how to match them across faults. Sometimes, faults were missed altogether. All these uncertainties had meant that Sheffels was, at best, going to be able to estimate only the smallest amount of squeezing that could possibly account for the folds and faults, though the actual amount of squeezing could have been much greater. To be precise, she showed that the crumpled layers of rock, which today span a region slightly less than 700 kilometers wide in the Bolivian Andes, were originally laid down flat across at least 900 kilometers. To achieve this, one edge of the pile of rock layers had been pushed horizontally toward the other edge at least 210 kilometers. This distance is often referred to as the amount of shortening of the strata. Converted into thickening of the crust and building the mountains, Sheffels's estimate of the shortening was enough to push up a mountain range, at the very least, to half the height of the present Bolivian Andes.

By now, our Andean project had come up with a much more detailed history of the mountains, and it was clear that we should use the Cretaceous layers, laid down just over sixty-five million years ago, as our once flat sheets or layers of rock. These had been deposited at the bottom of a vast inland sea or lake, not long before the present mountains started to emerge. Armed with this information, the trick is to draw a cross-section through the layers on a long strip of graph paper, putting together all the available geological observations—we also had those precious geological maps given to us by Werner Gutentag in Cochabamba—and showing how the layers have been broken and twisted today. But there is an art in this—you need to have a feel for the shapes of the rock layers as they weave and wind their way in the bedrock, and also you must make sure that they really can be reconstructed into a series of originally flat sheets. A major difficulty is working out what happens deep in the crust where you cannot see the rocks. And so there is always going to be a certain amount of trial and error in this work, plus the need for some basic geological intuition, but gradually a picture builds up on the long strip of graph paper of the huge displacements in the crust. With a deft stroke of the pencil, you can conjure up many kilometers of movement without having to stir from your desk.

Before I could have full confidence in any cross-section drawn through the Bolivian Andes, I had to deal with some confusing observations. In the early 1990s a team of French geologists, lead by Thierry Sempere and Patrice Baby, published a series of fascinating papers describing an extraordinary and uncharacteristic feature in the rocks on the western margin of the Eastern Cordillera. They described a large mass of rock, several tens of kilometers across, which had somehow become detached from the underlying bedrock—forming what geologists call a nappe—sliding horizontally many kilometers along a large and gently inclined fault. Their publications contained geological maps of this feature, near the village of Calasaya, and numerous cross-sections and diagrams illustrating the Calasaya nappe in all its many guises. Nowhere else in this part of the Andes had anybody found anything quite like this, and, I have to confess, I was very skeptical. So finally, in 1993, Lorcan and I went to have a look. We scrambled up steep hillsides and walked out particular rock units, trying to find the nappe. And everywhere we looked we found something different from what the French geologists had described—either steeply inclined fault surfaces, where they saw gently inclined ones, or natural depositional contacts between rock units, where they described faults. And so on. The Calasaya nappe was turning into a mirage or illusion—which is just what I think it is—and the closer we got to it, the further it seemed to recede into obscurity. You may ask why this mattered. Well, it did, because if the Calasaya nappe had really existed, it would have opened the door to the possibility of very large amounts of horizontal movement of the crust in this part of the Andes, confounding all the existing estimates of shortening.

Whereas the illusory Calasaya nappe had the potential to overestimate the shortening, another set of ideas about the Bolivian Andes had the potential to underestimate it. There has been a long tradition among geologists that wherever there are thick piles of red-beds, the crust has been pulled apart or stretched, creating a depression in the Earth's surface that is soon filled with sediment. The high plateau of the Altiplano contains an extraordinarily thick pile of red-beds, laid down by rivers flowing into the region at an earlier stage of its evolution. To some geologists it seemed natural to assume that these red-beds had accumulated along a rift in the crust, bounded by gigantic, steeply inclined faults tilted toward the center of the rift. Again, I was puzzled by this idea because it seemed so out of keeping with all the evidence for squeezing elsewhere. I devoted considerable effort to examining the now-tilted layers of red-beds, searching for the fault lines that bounded them. I was able to study not only the rocks themselves, but the results of the exploration activities of oil companies. They had been probing the ground using seismic vibrations, building up a picture of the underlying rock layers in what is called a seismic section. I was soon convinced that this region was not a rift at all, but had been squeezed throughout its evolution, even while the red-beds themselves were being laid down—it was just that the edges of the Altiplano had been pushed up more than its center. Gradually, as one by one each misleading idea was weeded out, I felt I was coming closer to understanding the true nature of the movements in the crust beneath the mountains.

I have not been alone in my attempts to work out the shortening across the Bolivian Andes. Indeed, this has become almost a competition, as rival geologists compete with each other to produce the most accurate measurement. Perhaps the most prominent of these is Jonas Kley at the University of Hanover in Germany. The new estimates of shortening east of the volcanoes have risen slightly to around 300 to 350 kilometers, and possibly as much as 400 to 450 kilometers in all across the whole mountain range, though many uncertainties remain. Despite all of this, the estimates have remained stubbornly less than the shortening required to create all of the deep roots in the crust, discovered by Susan Beck's team, solely by squeezing. I should mention here something I have so far ignored. Not all the rocks that have been pushed up in the Andes are still there; they have been washed away by the powerful forces of erosion. Some of this detritus has been distributed within the mountain belt—for, example, ending up beneath the Altiplano, or accumulating on the margins of the eastern foothills. But some has been carried by rivers much farther afield, as far as the Pacific and Atlantic Oceans. Ignoring all this will underestimate the volume of crust involved in the squeezing. Fortunately, because the Cretaceous lake-beds, deposited before the mountains first began to rise, are preserved throughout much of the Bolivian Andes, one can place limits on what the volume of this "missing" rock might be. Though a sizable amount, it is still a small fraction of the total volume of crust we are considering. We can, therefore, fairly easily calculate that it would take 500 kilometers or so of shortening to fully produce the roots of the crust; the measured squeezing is enough to make about four-fifths of their volume.

So, smoothing out the creases in the rock layers has turned out to be the key to understanding how the crust has been thickened beneath the Andes, pushing up the mountain range. Not only has it shown us the great importance of horizontal compression, squeezing the layers, but it has pointed out the need for another mechanism to create the remaining fifth or so of the volume of the mountain's crustal roots. It turns out that most of this "missing" volume lies beneath the high volcanoes in the Western Cordillera, something that became clear to us only when we turned our Andean project to studying this part of the Andes and provided, at last, a role for the volcanic activity that had long impressed both miners and geologists in the Andes—this is a subject for a later chapter. We can be confident, though, that for vast tracks of the Bolivian Andes, east of the volcanoes, squashing rocks really is the way to make these mountains.

Measuring the Height of Ancient Mountains

By the mid 1990s it seemed that the general problem of the creation of the Andes had been largely resolved. For the most part, the mountains had been squeezed horizontally. To many geologists this seemed to be the end of the matter. They had the big picture and were happy to let Andean "specialists" sort out what they considered to be a few relatively minor details. But often in science, the details can prove to be more interesting than the big picture and, on occasion, can cause that picture to change. Being one of those "specialists," I was not quite content. I certainly felt that we now had a much clearer idea of the thickness of the crust, though, it must be admitted, for only one part of the Andes. I realized, however, that if I truly wanted to understand the way these mountains rose, unraveling what would probably turn out to be a complex sequence of events, I would need to know much more. Ideally, I would need to know about the thickness of the crust at all the intermediate stages of the mountains' growth. This seemed, at first, an impossible undertaking. But there was a glimmer of hope—a potentially simple way to peer back through time. If I knew how high the mountains were at any point in the past, then I could make use of Airy's iceberg principle to work out the corresponding thickness of the crust as well, in effect monitoring the growth of the mountain's roots through time. This could be checked to see if it matched with the history, recorded in the rocks, of squeezing of the crust.

By studying the rocks, we can be fairly precise about when the mountain ranges started to form—some forty million years ago in Bolivia. And, of course, the present heights of the mountains are clear—about four kilometers or so in the Bolivian Altiplano. However, I knew very little about the heights of the mountains between these two dates. This was a bit like monitoring the growth of a child into adulthood without any statistics on his or her height, except as a newborn baby and an adult. Some children shoot up rapidly, virtually reaching their full height before their early teens. Others are late developers, showing a sudden spurt of growth much later on. In either case, the story of their growth gives a unique picture of their development. And so it is with mountains.

Unfortunately, working out the changes in height over time of a growing mountain range is a tough nut to crack. However, the 1990s were a particularly good time to be trying to find a suitable nutcracker. An exhilarating idea was being revived—first proposed nearly one hundred years earlier—that the rise of great mountain ranges can significantly change the planet's climate. And with the interest in the phenomenon of global warming, governments were prepared to fund research into testing this idea by trying to see if past changes in the global climate could be linked with mountain building. All this spurred on the search for a reliable geological altimeter, one that would be preserved in the rocks. Such an altimeter would have to be very different from the altimeters you can buy in a shop. The commercial ones measure the air pressure directly and use a standard relation between pressure and altitude to work out the height; the pressure decreases exponentially with height above sea level. For example, at an altitude of about 5,500 meters, near the summits of many of the peaks in the Bolivian Andes, the pressure is about half that at sea level.

It is not only the pressure that changes with height. As anybody who has climbed a mountain knows, it gets colder higher up as well. Of course, this temperature will vary from day to day, and be highly dependent on the weather. But, taken over a year, the temperature will be systematically colder with height. A typical decrease in the temperature of the atmosphere up to altitudes of around 10 km, known as the lapse rate, is about 5.5°C per kilometer. This means that at the summit of Mt. Everest, nearly nine kilometers above sea level, the air temperature could be as low as −50°C. As well as temperature and pressure, the atmosphere becomes much drier at higher elevations. You only have to visit the Bolivian Altiplano to experience this. I have found the air so dry that it actually affects the mucous membranes in my nose and becomes painful to breathe in. Conversely, the foothills of a mountain range are often exceptionally wet.

The most sensitive indicators of these changes in atmospheric conditions are plants and animals. The atmospheric pressure at altitudes above about 5,500 meters is too low for large animals, including humans, to survive for long periods. Botanists have studied the remarkable and systematic change in plants, and especially their leaves, up the side of a mountain. Lush jungle thrives in the wet foothills. Here, plants have large leaves to capture as much light as possible in the dim conditions of the jungle beneath the tree canopy. The high rainfall ensures that there is no danger of the plant losing moisture. In fact, they commonly have special adaptations to cope with the rain, including a pointed drip-tip to shed water. At higher elevations, the leaves become progressively smaller and more waxy, to conserve moisture, and the leaf changes from one with a complicated indented outline, at moderate altitudes, to simple, smooth shapes at high altitudes. Above about 2,000 meters, it is generally too cold and dry for large trees to survive naturally. The vegetation becomes more shrubby and smaller. Vascular plants with woody stems become almost nonexistent above 5,000 meters.

Occasionally, leaves are fossilized, and their shapes are almost perfectly preserved in the geological record. This can happen in a number of ways. An ash cloud from a large explosive volcanic eruption can spread out hundreds of kilometers from the volcano. Eventually, the rock fragments fall back to the ground, burying the landscape and any plants in a layer of fine ash. Alternatively, leaves falling off a tree or bush may sink into a lake. Here, on the lake bed, the leaf may become entombed in mud. Subsequent ash falls, or mud and sand layers, build up on top, compressing the

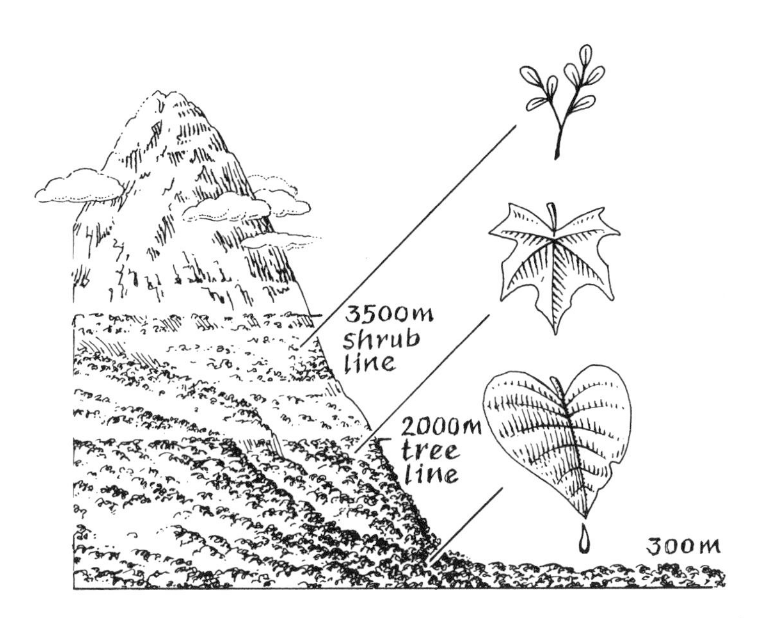

Temperature and humidity generally decrease with elevation, and plants have adapted by modifying the shapes and sizes of their leaves. Geologists can use a fossil leaf to estimate how far up the side of a mountain the original plant grew.

underlying layers. The shape of the leaf can leave its impression in the volcanic ash or mud, preserved in the rock for millions of years. A geologist has only to find the layer and split it open to once again reveal the leaf. Sometimes, the fossilized leaf may even have retained its original color, though this soon fades away on exposure to the air. But enough of the leaf remains that it can be used as a sort of geological altimeter.

Fossil Leaves in the Andes

In the 1920s, Johns Hopkins University in Maryland mounted a series of expeditions to the Bolivian Andes to collect fossils. One of the expedition members, John Berry, was looking for fossilized plants. He had heard rumors that Bolivian miners, as they prized open the layers of rock, were finding the remains of leaves. Fossil leaves were particularly abundant in two mines, both situated thousands of meters above the present tree line in the high Andes. I had already visited one of the mines on my first trip in Bolivia with Raul Carrasco. This was the copper mine of Corocoro in the Altiplano, at an altitude of about 4,000 meters. Here, the fossil leaves were found in mud layers between slabs of red sandstone. The other mine was probably the most famous mine in the world—the Cerro Rico, or "Mountain of Riches," in Potosi. At an altitude of nearly 4,500 meters, the leaves were delicately preserved in layers of volcanic ash on the side of the mountain. Berry made a large collection of rocks containing these leaves and shipped them back to Maryland.

In 1938 Berry published a detailed account of the fossil leaves he had collected in Bolivia. He had found leaves from several tens of plant species. Botanists call such a collection a flora. Berry concluded that the Bolivian flora from both mines had originally flourished at a much lower altitude. By comparing the leaves with the most closely related plants living today, he concluded that the altitude at which the flora had grown was not much more than 2,000 meters—this is over 2,000 meters lower than the altitude that the fossils are found today. Berry acknowledged that his estimates were only very approximate, but they were highly suggestive of substantial uplift. However, the real significance of these results depended heavily on the age of the rocks in which the fossil leaves had been found. The younger the rocks, the more interesting it would be. Berry, himself, did not know the age, though he guessed that the leaves were not more than a few tens of millions of years old. Our own and other workers' efforts on the dating of volcanic ash, from above and below the leaf-containing layers, have now shown that the fossil floras from both Corocoro and Potosi are somewhere between fourteen and eighteen million years old. And, unlike Berry, we also knew that the Bolivian Andes had first started to rise some forty million years ago. So it was beginning to look as though these mountains were a bit like the child that was a late developer, rocketing up in his or her teens.

Berry's work was largely ignored, and his collections lay forgotten in American museums. However, in the last ten years, with the renewed interest in determining the height of ancient mountains, the technique of studying the leaves has been developed considerably since Berry's day. Rather than compare the fossil floras with the nearest living relatives, it is possible to analyze the shapes of the leaves in detail, looking for features that are associated with evolutionary adaptations to particular altitudes. This method was developed by an American paleontologist, Jack Wolfe, who had used it to work out the original elevation of a fossil flora from the Colorado Plateau in Arizona. When I first found out about Berry's work—by chance, as a reference in another obscure article written in the 1950s—an American geologist, Kate Gregory, was already visiting Bolivia to collect more fossil leaves. She found floras in the Altiplano and Eastern Cordillera that were younger than Berry's leaves, dated to around eleven million years ago. Using Jack Wolfe's more quantitative approach, Gregory showed that her new Bolivian fossil floras could have flourished in a climate 10°C to 13°C warmer than today, such as that found at elevations as much as three thousand meters or more lower than where the floras are now.

The work on fossil leaves, if taken at face value, is convincing proof that there are parts of the Bolivian Andes that have at least doubled their height during the last ten million years—a surprisingly short period of time. But we need to tread carefully here—this conclusion is based on a crucial assumption that I have so far avoided mentioning, which, if wrong, would fatally undermine the reliability of our floral altimeter. This is simply that climate itself, as part of global climate change, has not changed for other reasons, becoming much colder and drier at high altitudes. And even if the altimeter was right, one could not be sure of the full extent of the mountainous regions that had risen so rapidly, because the leaves had been found in only a few isolated places. Perhaps only relatively small parts of the mountain range had rocketed up like this? Or perhaps all of it had? To me this latter possibility would be truly amazing. I could imagine some small piece of crust rapidly squeezed up. But the whole mountain range? How could such a thing happen? Curiously enough, the answers to many of these questions were lying in the landscape of Bolivia, clearly visible to anybody who happened to pass through.

Flat Plains in the Mountains

When we first started searching for clues to the growth of the Bolivian Andes, we where unaware of either Berry's or Gregory's work. We had taken a rather different approach, studying the deep gorges and valleys carved out by the main rivers in the Eastern Cordillera on their way to the lowland plains. What we discovered can be clearly seen on the journey by road from the old capital of Bolivia, Sucre, to the ancient mining town of Potosi, to the southwest.

Sucre is situated in a gentle topographic bowl on a high upland plain at an altitude of about 3,200 meters above sea level. The white Spanish colonial buildings, dating mainly from the seventeenth and eighteenth centuries, reflect the harsh light in the dry and barren landscape. Unlike many of the towns in the rugged Eastern Cordillera, there is plenty of flat land in the vicinity of Sucre, and the town boasts an airport that has a long enough runway for big jets. The road to Potosi leaves from the southern end of the town and soon plunges down into a valley, descending nearly a thousand meters to the Pilcomayo River. The Pilcomayo forms one of the longest rivers in the Bolivian Andes, snaking its way across the Eastern Cordillera toward the lowlands, several hundred kilometers farther downstream.

In the early 1990s the road bridge across the Pilcomayo was still under construction, and it was necessary to ford the river. Once across the river, the road winds up the steep southern sides of the valley, eventually emerging on another high and wide flat plain. From here, if you look northward, you will see that you are at the same level as the plains around Sucre itself—with your head close to the ground, you would miss the thousand-meter-plus Pilcomayo gorge altogether. Continuing on the road, you reach a junction close to some hot springs. If you go south, you will eventually leave the upland plain and descend into another deep valley. If you go west, you will head across a level surface toward a range of rugged hills in the far distance. Behind these hills is the old town of Potosi, perched on the steep sides of a mountain close to the ancient silver mines. Today, the mines are nearly exhausted, and Potosi has the air of a place that is slightly down on its luck.

The high plains between Sucre and Potosi form a landscape that would astonish any geologist. It is almost as though some unseen hand has sliced off the mountain peaks with a sharp knife, leaving behind flat-topped stumps. There were clear signs of this where the steeply tilted layers of strata, traced upward in the underlying bedrock, intersected the flat-topped land surface. Here, they ended abruptly, truncated like the land surface itself. We soon found that it was not just between Sucre and Potosi that this unseen hand had been at work. It had hacked its way down the length of Bolivia, in a wide swath of country fifty kilometers across, from Cochabamba in the north to the border with Argentina five hundred kilometers farther south. We were seeing the results of quarrying of the bedrock on a gigantic scale, with the removal of millions of tons of overlying rock.

This strange landscape might not trouble travelers whose minds are more on where they are going to sleep the next night, or what they are going to have for dinner—more likely, they will merely be relieved that for at least some of their journey they are not grinding up or down steep hills. They have more pressing concerns. But to us, whose main reason for being in Bolivia was to worry about the landscape and underlying bedrock, the flat-topped mountains were a major topic of conversation. The three members of our group, Lorcan, Leonore, and myself, were in agreement that the only agents of excavation powerful enough to carry out such a large mining operation, beheading the mountains, are rivers. But it is not in the Andes themselves that you can see most readily how they might do this, but on the lowland plains to the east where the rivers leave the mountains and head across the interior of South America.

Here, the rivers meander across the landscape, making their way inexorably toward the sea. At each bend or meander, the water is thrown outward, moving faster to keep up with the rest of the flow. This makes the outside of a bend an unusually powerful part of the river, especially when the river is in flood, and tends to result in a progressive undermining of the river banks. This way, a river cuts sideways into the underlying bedrock. Gradually, over successive floods, the banks shift, amplifying the bend, and the river becomes ever more windy in its course. Sometimes it may even end up turning through nearly 180° forming tight loops called ox-bow bends. More fundamentally, this sideways shifting of the river is planing off the underlying bedrock into a more or less flat surface. The river carts away the rock detritus it has excavated, transporting it much farther downstream. The surface left behind is very slightly inclined in the downstream direction with the same gradient as the river itself—a typical gradient would be a fraction of a degree or one part in several hundred. Eventually, over a period of several hundred thousand years, as many rivers wriggle sideways all over the landscape like a mass of angry snakes, an extensive flat surface in the bedrock is created. It is as though the rivers are acting as a gigantic grinder.

There is one major problem with the idea of meandering lowland rivers carving out the flat-topped mountains that we had found in the Eastern Cordillera, a problem immediately obvious to us whenever we stood on top of one. This is simply that we are not in the lowlands, and the present large rivers in the mountains, such as the Pilcomayo and its tributaries, are nowhere near the level of the flat mountain tops, but lie well over a thousand meters lower at the bottoms of deep valleys. In fact, the rivers are nearly two thousand meters beneath the most easterly flat-topped mountains. We puzzled over this. I remember arguing vigorously with other members of our expedition, as we peered up from the bottom of the Pilcomayo gorge at the flat plains towering above us, refusing to believe the obvious explanation.

But there was no other way out. I had to accept that the flat-topped mountains were the isolated and uplifted remnants of a once far more extensive plain—we called it a peneplain. The peneplain must have been carved out when the mountains were much lower—in fact, between one and two kilometers lower—and the rivers carried away the product of their excavation down to the lowlands; there was no sign of the vast bulk of this detritus on the peneplain itself. As the land rose, the rivers started to cut predominantly downward, rather than sideways, through the underlying bedrock, becoming more and more deeply entrenched in their valleys. Having come to terms with this idea of uplift, so blatantly advertised in the landscape, we realized that we had found another altimeter to measure the height of the mountains in the past. If we could determine exactly when, and where, the early rivers had carved out the peneplain itself, tying in all this with the places where the fossil leaves had been found, we would be able to work out the story of the rise of the mountains.

Dating a Landscape

We first discovered the remnants of the peneplain on our first major expedition to Bolivia in 1990. At that time we did not really understand its full significance. We had, however, inadvertently found the first clues to when it might have been carved out of the landscape. In some places, where the road climbed up to the peneplain, it passed through a small road cutting. Here, a thin veneer of gravels—piles of rounded rock boulders—could be seen resting on top of the bedrock surface. These deposits seemed to be all that was left behind by the rivers that once flowed over the peneplain—perhaps the last rivers before the landscape was uplifted, and the rivers started cutting their deep valleys. We noticed thin layers of white material within the gravels. Closer examination showed that they were volcanic ash layers full of the minerals quartz and feldspar. These ash layers must have spewed out from one of the volcanoes much farther west. Wafted high into the atmosphere, the ash would have spread out as a giant dust cloud, blown by the wind far from the volcano before falling back to the ground within hours or days of the eruption.

Flat plains in the mountains. In the rugged Eastern Cordillera of Bolivia, the remnants of a peneplain are clear to see on the journey from the capital Sucre to the ancient mining town of Potosi. This peneplain must have been carved by rivers when the mountains were much lower, and it has now been uplifted nearly two thousand meters.

To our delight, the ash layers often contained small black flakes of a mineral called biotite, glinting in the sun. The intense black color and perfect crystal shapes of the biotite strongly suggested that this mineral had crystallized from molten rock in the volcano. Biotite is also rich in the element potassium—a fresh crystal of biotite should contain about 7 percent potassium by weight. We were confident, therefore, that we could date the volcanic eruption using the radioactive decay of potassium in the biotite crystals as our clock. We carefully collected samples of the ash layers, so that we could extract enough biotite for the dating analysis. Gradually, we began to accumulate dates for volcanic eruptions that had thrown volcanic ash onto the peneplain. But in late 1993, after our fourth field season, I received a jolt when I read a paper by Timothy Gubbels and two colleagues from Cornell and Queen's universities in the United States and Canada. They had already worked out the story of the peneplain, mainly from studying its remnants farther south nearer the border with Argentina. And they had seen the significance of this feature for uplift of the mountains, as well as arriving at a timetable for its rise, based on their new dates for various volcanic eruptions.

Both myself and Lorcan—with whom I had been closely collaborating on the peneplain enterprise—were simultaneously disappointed and satisfied by their results. It showed that we had been on the right track, but we had been beaten to publishing our conclusions. However, the Gubbels paper was very short—only four pages long—with little detail or documentation. It was the outlines of a case, but not yet the full brief. We decided that we would now bide our time, completing the job. We would take the peneplain apart—scientifically speaking—and really try to understand its origin and subsequent evolution. In fact, I still have a student doing this, nearly ten years later! And so I make no apology for the fact that the rest of my account is our view—an Oxford perspective—of things, showing how we tried to tease out the secrets of the landscape in this part of the Andes.

We had all clearly hit on the evidence for an extraordinary geological phenomenon. Putting all the new dates together, it became clear that the peneplain appeared in the landscape at the end of the Miocene period of geological time, some ten million years ago—not long before our hominid ancestors diverged from the Great Apes—carved out by a large system of meandering rivers in the Eastern Cordillera. This age is tightly constrained because the youngest strata in the bedrock beneath the peneplain are about twelve million years old. And the oldest coverings of volcanic ash that settled out on top of the peneplain are about nine million years old. Our own dates for these mantling volcanic ash layers showed that they were mostly erupted between three and six million years ago. Since then, the peneplain, still nearly flat and preserved almost intact—except for the inroads of the deep valleys where younger rivers had eaten down into it—had been raised up somewhere between one and two thousand meters as part of the growth of the Bolivian Andes. However, as we expanded our dating work, so the region that had been uplifted seemed to get larger as well.

We had begun to examine and date volcanic rocks from other parts of the Eastern Cordillera. Some of these lie immediately west of the mines of Potosi, where the high peaks are the eroded stumps of volcanoes, and most of the surrounding landscape is underlain by thick layers of volcanic rock called ignimbrite. These ignimbrites—the Los Frailes ignimbrites—are full of fresh crystals of quartz, feldspar, and biotite, welded together into a compact, dense rock. Occasionally exotic lumps of bedrock are embedded in the layers. All this stands as stark testimony to volcanic eruptions on a scale that has never been witnessed in historical times. The ash, together with lumps of bedrock, was blasted out of a volcano as a red-hot rock and gas mixture, reaching temperatures of hundreds of degrees centigrade. This vast, deadly cloud swept out over the landscape, hugging the ground and traveling at speeds well over a hundred kilometers an hour. Eventually, as the ash fragments settled out and cooled, a hard volcanic rock formed.

Satellite pictures of the ignimbrites clearly revealed the full extent of the volcanic eruptions—they had spread out in a roughly circular region nearly a hundred kilometers across. Another, similar volcanic center could be found about a hundred kilometers farther north. And the eruptions were preserved more or less intact, without faulting or tilting of the ash layers. It was clear that they had erupted in the very period when the peneplain itself was created. In fact, not far from Potosi, we traced one of these eruptions close to the main peneplain surface. In this case, it was hard to conceive how the peneplain could have risen without bringing up the ignimbrites with it. And the Cerro Rico, where the American paleontologist John Berry had first discovered his fossil leaves, was right among these ignimbrites. So here, at last, we had a link between the two different altimeters.

When two independent methods agree, scientists always feel much more confident that they are on the right track. The two altimeters were individually only approximate ways of detecting the rise of the mountains. But together they were very powerful indeed, revealing the full significance of our discoveries. We were talking about the equivalent of lifting a region the size of England and Wales, or half of California, from sea level to the level of the top of the Grand Canyon, and much higher than any mountain in the British Isles. So far, I had convinced myself that the rise of the Bolivian Andes could largely be explained by horizontal movements in the crust—the evidence for this could be seen almost everywhere in the folding and faulting of the bedrock. These movements would squeeze and thicken the crust, creating deep roots, and, according to Airy's iceberg model, lead to an inevitable rise of the land surface. This should also apply to our new discoveries. Indeed, using Airy's iceberg model, I should be able to work out exactly now much squeezing and thickening of the crust must have occurred to account for the observed uplift—I would expect the squeezing to be quite substantial. We were now about to face up to yet another problem, one that had, in fact, been there all along.

There was no sign in any of our new discoveries for this squeezing. The peneplain surface, together with its river gravels and volcanic ash layers, as well as the ignimbrites to the west, lay horizontal and apparently undisturbed. In other words, though there had clearly been squeezing of the bedrock before the creation of the peneplain, there had been none subsequently. So how could the peneplain have risen? Had we got it all wrong after all? My experience in geological research is that if you have a problem, make sure that you have examined all the evidence—there is usually enough to solve it, but only just enough, and you have to look everywhere to find what you need. There were parts of the mountain range that I had so far ignored, except for a reconnaissance visit early in our project. These were the forested regions that lay even farther east, along the foothills of the Bolivian Andes. It turned out that this was where the answers to my questions were to be found.

Tales from the Foothills

The foothills of the Bolivian Andes, or the Sub-Andes, form a very different world from the bare, dry hills of the high Andean ranges farther west. The lower ranges of the Sub-Andes rise up out of the Amazon jungle. Lush vegetation creeps up the hills, cloaking the rocks with a green mantle of plants. Here, there are several meters of rain each year, and the hills are frequently shrouded in clouds. This is not a friendly landscape to penetrate. The threat of malaria is everywhere. There are poisonous snakes and insects ready to bite if disturbed. And the vegetation has teeth of its own—a spiny undergrowth that rips your clothes and skin. All this in a hot and humid atmosphere that soon has you bathed in sweat after the slightest exertion. You need to be a very determined geologist if you want to study the rocks here.

Viewed from space, the Sub-Andes reveal many of their geological secrets. The layers of rock strata have been worn away by slightly different amounts. More resistant layers stand proud as ridges, while the weaker layers form grooves between the ridges. On the ground, these subtle differences are obscured by trees. But from the air they are clear to see and can be easily traced out in the satellite pictures. These show that the strata have been folded into a series of great folds—synclines and anticlines in the jargon of geologists. The same distinctive sequences of rock layers appear many times in the landscape, not only folded, but thrust up and repeated by great faults that cut through the rock layers.

The great folds and faults in the Sub-Andes are like gigantic broken corrugations in the crust. These run parallel to the general trend of the hills and valleys, following the line of the edge of the mountain range. Farther east lie the plains of the Amazon Basin. The Bolivian national oil company, YPFB, has been investigating these regions for many years as they search for new reserves of oil. I had seen some of the results of their work during my first trip to Bolivia in 1989, when I visited the YPFB offices on the lowland plains in Santa Cruz. Their exploration activities have left a legacy of a myriad delicate lines imprinted in the landscape where long thin strips of forest have been cleared away for the seismic crews and their equipment. By setting off explosions and recording the vibrations with detectors arranged along these lines, the oil companies have produced a network of seismic sections.

The seismic sections reveal the shape of the rock layers, not just at the surface, but also deep in the crust. The mountain front itself is a major fault in the crust. To the east, beneath the lowland plains of the Amazon jungle, the strata are virtually horizontal. But a sideways view of the folds and faults, farther west, within the rock layers beneath the Sub-Andes themselves, conjures up an image of wavelike mountains. Now that I have been allowed to see some of the seismic sections—the oil industry usually keeps them well guarded—I find that whenever I look toward these mountains from the lowland plains, I can't help thinking that I am on a beach pounded by huge waves of rock. The first range of hills seems to be where the sea, flowing up the beach as the wave breaks, meets the land. Each range of hills in the distance is like a wave farther out to sea, inexorably moving closer to the seashore. The vast ranges of the Andes seem to be crashing at my feet.

The fundamental cause of my strange vision of the Andes is the intense squeezing of the crust beneath the Sub-Andes, distorting the strata. The amount of squeezing that must have occurred can be measured, in effect stretching out the contorted layers back to their original flat orientation—I get my class of undergraduates to go through this exercise as part of their course work. They are always surprised to discover that the mountains must have been pushed horizontally, toward the east, more than a hundred kilometers. So when did this squeezing or shortening take place? The earthquakes in this region suggest that it is going on today. The most easterly fault that separates the first low range of hills from the flat plains is certainly active, moving intermittently during earthquakes. In this case, the question really becomes, when did the movements start?

The answer depends on the age of the rock layers. Thick piles of red-beds lie tilted and faulted in the cores of the great, gutterlike corrugations that follow the valleys in the Sub-Andes, exposed in the banks of rivers. These were laid down by much older rivers that once flowed off the flanks of the Andes, rivers that were certainly flowing about twenty-four million years ago, because a volcanic ash layer near the bottom of the pile of sandstone beds has been dated to this time. The red-beds lying a few thousand meters higher up in the contorted pile must be several million years younger. And it is clear that these strata started to be squashed together even later on, and this squashing is continuing today. It is hard to escape the conclusion that these movements were taking place at the same time—during the last ten million years—when the peneplain and ignimbrites farther west, in the Eastern Cordillera, were uplifted as a solid block. The logic of all this is relentless; somehow, the squeezing in the Sub-Andes is connected with the uplift of the peneplain and ignimbrites. And this is where the measurements that I spoke about at the beginning of this chapter, those measurements of *g*—the acceleration due to gravity—laboriously made throughout the Bolivian Andes, come into their own, providing a surprising piece of evidence for what this connection might be.

A Shank in the Sole

I have already told how gravity observations in the Andes showed us that one or another of the two main models for the nature of the crust beneath mountains—Airy's or Pratt's—is basically right, but they failed to determine which one. If Airy was right, then it was a simple matter to calculate what the thickness of the crust should be using his iceberg principle. As a general rule of thumb, given the best estimates of the average density of the crust and mantle, the base of the crust should extend a farther seven kilometers deeper into the Earth for every extra kilometer that the mountain range rises up above the surface. However, it was not until detailed measurements of the thickness of the crust using seismic vibrations had been made that it was possible to check this. It immediately became clear that it was Airy's model after all, in which the crust floats like an iceberg on the underlying fluidlike mantle, with deep roots in the crust beneath the high Andes, that fitted the seismic results best.

When it finally became apparent that we were mainly dealing with an Airy model for the crust beneath the Andes, then it was time to look at the gravity data again, but now much more closely, searching for any discrepancies. It is a relatively simple matter—especially if you have a powerful computer—to work out what the pull of gravity should be beneath any part of the Andes if Airy's model is right. In this case, we can compare this theoretical calculation with the actual measurements of gravity. Any difference between our calculation and the observations will tell us that Airy's model was not working. These discrepancies, or anomalies, as geophysicists prefer to call them, turned out to be of crucial importance to our work.

In 1993 I finally persuaded Tony Watts, a colleague at Oxford who is an expert on gravity, to look for these anomalies by making the appropriate calculations. We found that there were, indeed, significant discrepancies in the eastern foothills—in the Sub-Andes—and also the lowlands farther east, and they could be traced for hundreds of kilometers along the length of the mountain range. We can divide these up into a positive anomaly, where the pull of gravity is greater than it should be according to Airy's model, and a negative anomaly, where it is less. Tony had spent many years looking at just such a pattern, observed in gravity measurements from many parts of the world. As we contemplated the data in his office, displayed on a computer screen, Tony traced out with his finger the up-and-down shape of a plot of some of the Bolivian gravity measurements. He turned to me with a smile on his face: "Simon, you know what we have got here?" "No," I said truthfully. "It's flexure, not pure Airy at all." Tony had immediately realized that Airy's model needed to be slightly modified to explain the gravity data. Airy had assumed that the crust will sink down in direct response to its local weight—this weight increases as the crust becomes thicker—like a laden ship floating on the underlying "sea" of mantle. In reality, the relatively "strong" outer shell of the Earth behaves more like a flexible diving board or plank, warping or bending over a wide region under the load. It is as though there were a metal shank or stiffener buried deep inside, like that in the sole of a hiking boot, imparting this rigidity or flexure.

It is not hard to imagine how the weight of the mountains on the eastern margin of the Bolivian Andes might push down on this springy plank. As a result, the plank will bend, though its strength will help to hold up the mountains, while at the same time spreading their weight over a much wider region, extending into the lowlands. Because of this, the base of the crust beneath the mountains will be held up higher than it otherwise would be if the crust were allowed to float free, showing up in the gravity measurements as a positive anomaly. However, the base of the crust below the lowland plains, immediately east of the Andes, will now be pushed down deeper than normal—here, gravity measurements will reveal a negative anomaly. Oil companies had already found direct evidence of this pushing down of the crust in the lowland plains of the Amazon jungle. A wedge-shaped trough or depression, up to five thousand meters deep, has been created in the Earth's surface, now filled up with river detritus washed down from the mountains as the Andes have grown. The wedge of detritus has turned out to be crucial for the economic life of Bolivia; this is where Bolivia's oil and gas reserves have ended up, forced out of the adjacent mountain ranges during the squeezing of the rock strata.

Tony Watts had developed a way of using the gravity measurements, together with the general topography of the mountains, to estimate the springy behavior of the South American continent.

This can be measured in terms of the equivalent thickness of a springy elastic plate or stiffener. And just as with the metal shank in the hiking boot, the greater the thickness of the stiffener, the less flexible it will be—in reality, of course, this stiffener or shank does not actually exist as a separate entity in the Earth but is merely a representation of the springy strength properties in the South American continent. In fact, Peter Molnar and Helene Lyon-Caen at MIT, and Dean Whitman at Cornell, had already estimated this springiness beneath a part of the Andes in southern Bolivia and northernmost Argentina. However, Tony had managed to get hold of all the available gravity measurements in South America, so we could now expand on these results and work out the shape of the shank in the sole of the whole of South America. Subsequently, two of our students continued the work as part of their doctoral theses.

It fell out quite naturally from the gravity measurements, without any need for fudging, that South America behaves as though it contains two separate and rather irregularly shaped shanks or stiffeners. Rather satisfyingly, the centers of these two shanks coincide with major geological features of South America. There are two ancient portions of the continent, long known about by South American geologists and usually referred to as shield areas. The Amazon River flows between them as it traverses the breadth of South America. One is in the highlands of Guyana. The other, called the Brazilian Shield, is in virtually the center of South America, covering parts of Brazil, Bolivia, and Paraguay. Both these shield areas have lain quietly—almost asleep—for substantial periods of geological time, while all around huge upheavals have created the Andes farther west, the Caribbean to the north, and the Atlantic ocean in the east. In fact, British Survey geologists investigating the Brazilian Shield in the mid-1970s had shown that nothing geologically worth speaking about had happened here for the past six hundred million years. By imbuing the inactivity of the Brazilian Shield with new significance, we were about to wake it up!

In the Earth, lack of geological activity means plenty of time for the hot rocks beneath the surface to become cooler. And as the rocks cool, they usually become stronger—I will talk more about

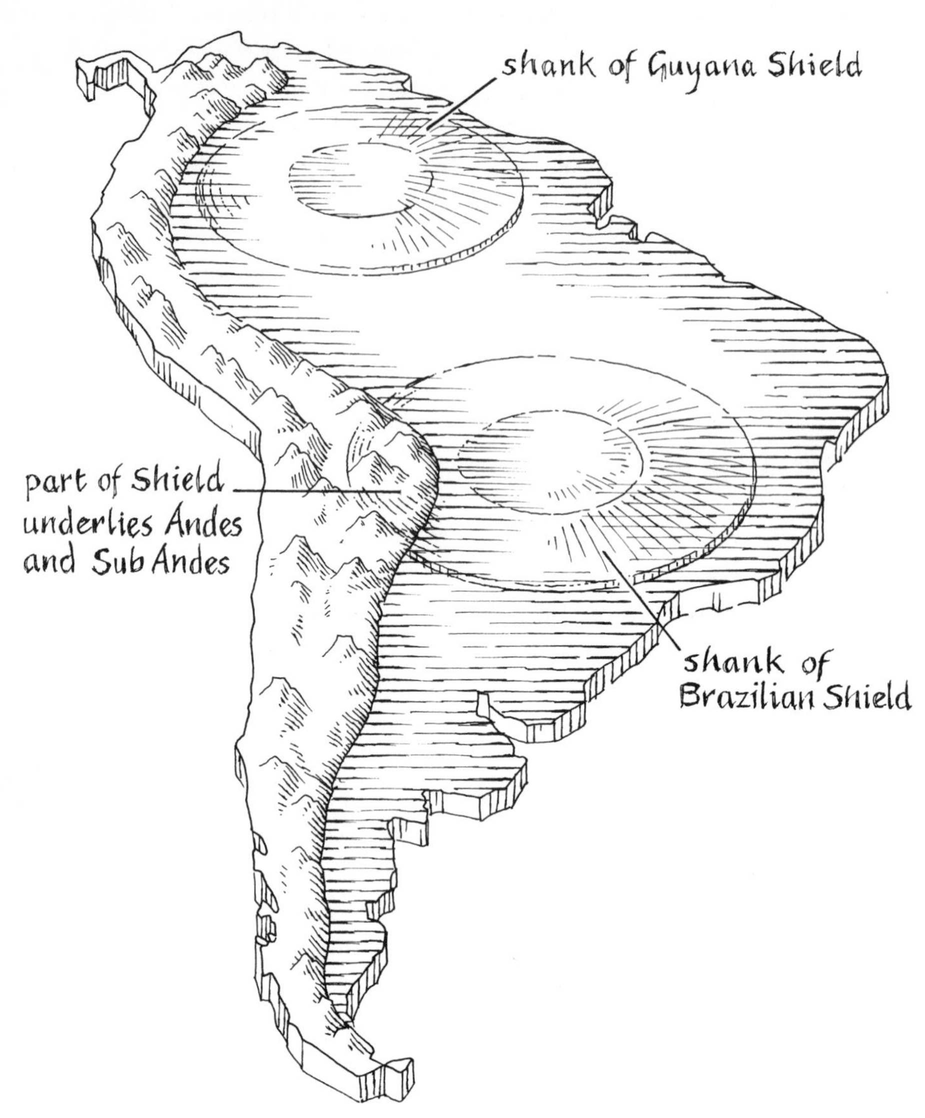

A shank in the sole of South America. Continents can be bent or flexed by the weight of a high mountain range like the Andes. Estimates of the stiffness or resistance to flexing suggest that South America contains two strong portions or shanks in the geologically ancient Brazilian and Guyana shields.

this in later chapters. So it was no surprise to us that the shanks or stiffeners turned out to be thickest—over fifty kilometers thick—beneath the centers of the shields. But these shanks, though slightly thinner, also extend beneath the Andes. Thus, the foothills of the Andes in Bolivia—the Sub-Andes—sit above part of the shank in the Brazilian Shield. It is this curious image of the shank or stiffener that helps us solve, at last, our original problem: how squeezing in the Sub-Andes is related to the rise of the Andes farther west, in the Eastern Cordillera.

Walking the Plank

Let me quickly review where I have got to in my story. I have shown that large tracts of the Bolivian Andes, in the Eastern Cordillera, must have risen up vertically well over a thousand meters in the last ten million years, almost as a sold block, without any sign of horizontal squeezing. However, the Sub-Andes, slightly farther east, were being intensely squeezed at this very time. And even farther east, the Brazilian Shield lay undisturbed, strengthened by a very thick stiffener or shank. We could summarize all of this by saying that the Eastern Cordillera had been pushed toward the Brazilian Shield, squashing the intervening Sub-Andes and somehow bobbing up as well. So far, so good. Continuing with our analysis of the gravity measurements, we find that we can explain the gravity anomalies only if a relatively thick stiffener lies beneath the Sub-Andes. Somehow, we suddenly seem to have got ourselves into a bit of a tangle, because it makes no sense for such a stiffener to be where the crust is being intensely squeezed and broken—the presence of this stiffener should "stiffen" the crust.

Yet again in the long and convoluted tale I have tried to tell in this chapter, we reach a stumbling block—an apparent paradox. But it dawned on me that I had already seen the solution. What about the gently inclined fault that I described earlier in this book—the one that was discovered in Scotland by those pioneering geologists, Benjamin Peach and John Horne, over a hundred years ago: a gigantic fault that had first opened the eyes of European geologists to the enormous scale of the horizontal movements in the Earth's crust? Or the fault that Henry Cadell observed soon afterward in his experiments, which seemed to be an inevitable consequence of trying to squeeze layers of clay in a vice? So if we think of the squeezing in the Sub-Andes to be rather like wrinkling up and sliding a carpet over a solid floor, where the floor represents the stiffener or shank, then all becomes clear. A sideways view of the crumpled rock layers in the Bolivian Sub-Andes would show them to be in a long, thin wedge, resting on top of the planklike Brazilian Shield. And the base of this wedge, separating it from the underlying Brazilian Shield, would be the gently inclined fault.

The long trail of geological clues has finally led us to a gigantic fault in the crust beneath the Sub-Andes. Except for its tip, where it emerges at the surface along the very edge of the mountains, the fault lies buried and completely hidden from view. Yet movement on this fault, as the Sub-Andes are squeezed, must be the explanation for the rise of the peneplain and ignimbrites. If the fault extends farther west, beneath the peneplain and ignimbrites, then it is not a big step to imagine that these regions are also sliding up the fault over the Brazilian Shield. As they slide up, so they rise. And the amount of squeezing observed in the Sub-Andes—a hundred kilometers or so—records how far the peneplain and ignimbrites, following on behind, have traveled up the fault in the last ten million years. It turns out that this amount of displacement on the fault, taking account of the tendency for the Brazilian Shield to sink under the weight of the mountains, is more or less the right amount to explain the observed uplift—about two kilometers—of the peneplain and ignimbrites.

Similar faults have been found, not only in Scotland, but also in other parts of the world, exposed in the rocks of the European Alps and the North American Rockies. But, as in Scotland, these faults are long dead, and their deep interiors are exposed at the surface only because the process of erosion has worn the mountains away. However, the discovery in Bolivia, as well as that recently made in the Himalayas, shows what these faults were really like when they

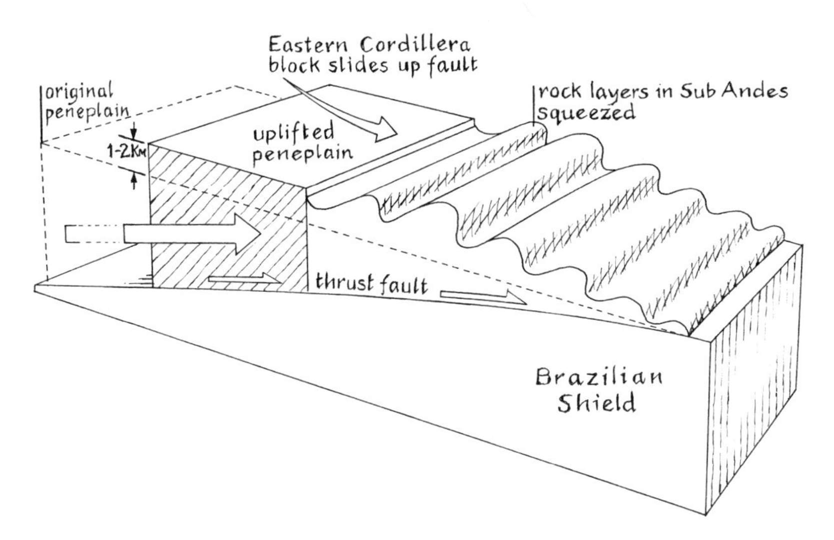

A great fault must lie hidden beneath the Bolivian Andes, emerging only at the extreme eastern edge of the mountains. As the rock layers in the Sub-Andes were squeezed, the rocks farther west slid up this fault, raising the landscape in the Eastern Cordillera.

were in action, pushing up the mountains and thickening the crust. Thus, by a curious twist of geological research, my work in the Andes had helped me to understand much better the significance of the geology in a part of Scotland where I had long been taking parties of undergraduates on field trips—you can be sure that the students got to know about this as well!

In the last ten million years, ever since the earliest hominids diverged from the apes and embarked on a course of evolution toward humankind, the Andes have been caught like a pip. In the east, as we have now discovered, they have been sliding over the Brazilian Shield. But there is also another gigantic fault in the west—I will talk more about the evidence for this further on in my narrative—where the floor of the Pacific Ocean is slipping beneath the western margin of South America. And so the mountains themselves have been forced up to nearly double their original height in a gigantic wedge. This dramatic rise is just the latest phase in a longer history, over the last few tens of millions of years, of squeezing throughout most of the Andes; a squeezing that has, in general, thickened the crust in proportion to the height of the mountains. Despite the modifications to his theory, the surveyor Airy had won over the mathematician Pratt, and his iceberg principle can, in essence, be assumed—to quote Airy—with "perfect correctness."

As the mountains rose, they changed shape in yet another extraordinary way. This is the subject of the next chapter.

✣ ✣ ✣ ✣ ✣

Rough Roads

When I first visited Bolivia, in 1989, over nine-tenths of the road system consisted of rough dirt roads, with only short lengths of sealed tarmac highways on the outskirts of a few major towns. The main dirt roads were continually being damaged by heavy articulated trucks, reduced to corrugated and potholed trails that traversed the landscape. These arteries of the road system linked together an intricate network of rough sandy or rocky tracks. Navigating any of these was feasible only in a truck or four-wheel drive vehicle. This left the main cities as virtual islands, each with their own unique population of ordinary cars that never left the city limits. Traveling by road between major towns was very slow, and one could not expect to average much more than about 30 kilometers per hour (20 miles per hour). A journey became a test of endurance, as the vehicle lurched from one pothole to another, wallowing in thick mud or powdery dust.

In the early days of our field work, gas stations were rare, especially in the high Andes of southern Bolivia where, even today, one gas station in Uyuni still serves a region the size of England. Visiting this station was always a critical moment. If Uyuni was out of gasoline—a real possibility—our travel plans would be severely curtailed. When there was gas, it was a slow business filling up. Nobody would dream of tanking up with less than a few hundred liters. And the main gas tank was only one of several that trucks or four-wheel drive vehicles carried. I never liked leaving Uyuni with much less than two hundred liters of fuel. Even then, on long trips among the giant chain of volcanoes in the Western Cordillera, it was necessary to scrounge gas from mining camps.

Uyuni did offer one luxury. It was on the edge of the Salar de Uyuni, the largest salt lake in the world. During the dry season, in the Southern Hemisphere winter months, the Salar was a gigantic, smooth plain of salt. After the rough roads of the rest of the Altiplano, the roads, or alignments as they are called, on the Salar seemed like motorways. Wherever possible one would search for a route across the Salar when trying to reach some distant point in the Altiplano. It was just a matter of deciding on the compass bearing, then driving in a straight line at full speed, the ground beneath you a brilliant white, and the sky above a deep blue. The longest journey of this sort that I have made was over two hundred kilometers. During the wet season the Salar fills up with a layer of water a few tens of centimeters deep. If you are confident, you can still cross the Salar in these conditions; far from dry land, the car seems to skim over the surface of the water like a speedboat.

Clues to long-lost mountains. The contorted strata on the right once formed the roots to a mountain range that existed over two hundred million years ago. The mountains were eventually worn away and covered by beds of sandstone and siltstone about sixty-five million years ago. These layers have now been tilted during the present phase of Andean mountain building.

The remains of a vast inland sea can be found in the Eastern Cordillera, near the village of Corocoro. Here, slightly over sixty-five million years ago, dinosaurs left their footprints in the mud. As slabs of hard rock, these layers are once again being walked over.

These pinnacles of red sandstone, near Camargo in the Eastern Cordillera, were deposited by rivers flowing eastward off the Andes as the mountains first started to rise, roughly forty million years ago.

A bedrock of pink and red, near the village of Salla. During the past few tens of millions of years, layer upon layer of red sandstone has been laid down in the Altiplano by rivers flowing westward off the rising mountains of the Eastern Cordillera.

Bands of volcanic ash, blasted out of a volcano some ten million years ago, now lie turned up on end in the Altiplano near Curahuara de Carangas.

Where has everybody gone? A donkey seems to be the only resident of this village in the Altiplano.

Llamas and sheep take to the high road in the Altiplano.

The wide gap between town and country. A campesino woman stands alone in the main square of La Paz during a public celebration.

Flat-topped mountains rise over a thousand meters above the bed of the Rio Pilcomayo in the Eastern Cordillera, looking as though they have been sliced off with a knife. In fact, these are the remnants of an extensive peneplain, carved out by rivers when the mountains were much lower.

The Cerro Rico—mountain of riches—dominates the city of Potosi in the Eastern Cordillera. During the Spanish colonial days, local miners extracted a large proportion of the world's silver from this mountain.

The Rio Pilcomayo finally reaches the lowland plains after a long journey through the Bolivian Andes.

The tree-cloaked foothills of the Andes rise abruptly out of the Amazon jungle on the banks of the Rio Grande.

These nearly horizontal strata on the eastern edge of the Andes are being squeezed by the same movements that have pushed up the mountains farther west — faults cut across the rocks, displacing the layers.

Not a good place to park! Our faithful Toyota Land Cruiser negotiates a precipitous road in the Eastern Cordillera.

The flattest place in South America lies at 3,600 meters above sea level in the Salar de Uyuni—a gigantic salt pan in the Altiplano. In the distance is Cerro Tunupa, an extinct volcano.

The road to the volcanoes. A typical desert track in the southern Bolivian Altiplano.

The bank of a sandy river in the Sub-Andes. This apparently peaceful scene was, in reality, disturbed by the whine of my rock drill. I was extracting a core of rock to be tested for its magnetic properties.

Car and geologists still asleep at sunrise—camping in the Eastern Cordillera near Otavi.

⁜ CHAPTER SEVEN ⁜

A Curvaceous Shape

Shape, ***n.*** Configuration, external form, total effect produced by thing's outlines.

Shape, ***v.*** Create, form, construct; model, fashion, bring into desired or definite figure or form.

Curvaceous, ***adj.*** Having many curves. (*OED*)

A DROPLET OF WATER swelled in the spout of the tap, becoming larger and heavier. Finally it detached itself, falling into the sink—"plink." And "plink" again. Would those taps never stop? "Don't turn off the taps," Leonore called. "Don't forget to leave the taps dripping," somebody else shouted for good measure. Those damn taps! We were in Ollague, right on the border between Chile and Bolivia at 3,700 meters above sea level, staying in a small prefab building lent to us by a local official. It was July, and the night-time air temperature was 20° below, cold enough for the water pipes to freeze up unless we kept the taps dripping. Or at least that was the theory. I made my way back to my bunkbed, past the rows of gasoline and water containers, stock-piled away from prying eyes in the narrow hall of our house. My rock drill, spattered with mud after a heavy day's use, lay nestled in its wooden box. Our equipment seemed so precious in this remote and barren place.

Ollague lies in the middle of a salt pan between two active volcanoes. It is both a border post, manned by the Chilean police, and a railway workshop for the main line between Bolivia and the coastal ports. A branch line also heads north out of the settlement, providing a link with several gold and silver mines. In the days of steam, large stockpiles of coal, and huge volumes of water in a metal tank on stilts, were stored here—now the water tank is derelict, with long daggers of ice hanging from it where water has been dripping from open rust wounds. I was here to collect samples of rock from the eroded stumps of a volcano that had erupted about eight million years ago—the Tres Monos volcano—coring the dark lavas with my rock drill. I hoped to use these samples to help solve a problem that has long intrigued geologists: why do the Andes have such a sinuous shape?

The shape of the Andes is clear to see on any map of the world. They snake their way down the western side of South America, following the curves of the coastline. One can liken this to the profile view of a crouching athlete: each twist or turn is a joint. The athlete's back is the coastline of Colombia and northern Ecuador, connected to the thigh of Peru at a hip joint in northernmost Peru. Farther south, at the latitudes of the border between Peru and Chile, one reaches the knee joint, where the coastline swings around into the calf of Chile. Finally, the southern end of South America is the ankle joint, attached to the foot of southern Patagonia. These swings or curves are mirrored, like the sinews of the athlete's back and leg, by the orientation of the tilted layers of rock in the Andes themselves, giving rise to geological bends or oroclines. The great swing in the mountain ranges of the Bolivian Andes, where we were mainly working, is called the Bolivian orocline. Cochabamba, our main base, is nestled right in the hinge of this bend.

Perhaps the curves were an ancient feature, following the lines of some earlier split with another continent hundreds of millions of years ago? In this case, in the terms of our analogy, the athlete had always been crouching. Alternatively, the athlete had once stood straight and upright and then subsequently crouched down: the curves were the product of later movements of the crust, perhaps during the growth of the Andes themselves. These movements, bending the mountain range into curves, could be detected if I had some way of working out changes of direction through time—a compass in the rock. Of course, a compass needle is just a bar magnet, and a magnet—or lode stone as it was called by early navigators—is just a special type of rock.

A Compass in the Rocks

Most rocks are magnetic, though they are usually only very weak magnets. Too weak, in fact, for you to notice unless you have a very sensitive way of detecting the magnetism. They acquire this magnetism when they form. For example, a lava flow erupted from a volcano may contain numerous minute crystals of a magnetic mineral called magnetite. Each crystal of magnetite behaves a bit like a bar magnet or compass needle. To begin with, the lava is too hot for the individual bar magnets to stay for long in any particular direction—the atoms at these temperatures are jiggling around too much—and so the magnets cancel each other out and the lava shows, overall, virtually no permanent magnetism. But when the lava eventually cools through a critical temperature, called the Curie temperature—named after Pierre Curie who discovered this effect nearly a hundred years ago—the microscopic "compasses" will finally settle down and begin to line up with the Earth's magnetic field. For magnetite, this temperature is 580°C; at lower temperatures the magnetite will retain a memory of the Earth's magnetic field, even if this field was somehow switched off. In other words, the rock will be permanently magnetized, acting as a bar magnet pointing toward magnetic north.

Now, imagine what would happen to the bar magnet if a mining company were to excavate the block of lava and lift it up with a crane. Suspended in the air, the rock might slowly spin around, turned by a twist in the cable or blown by the wind, before it was lowered onto a waiting truck. In this case, the direction of the magnetism locked into the lava—the orientation of the bar magnet or rock compass needle—will also rotate, no longer pointing in its original direction toward the magnetic North Pole. In fact, if I could measure the final direction in which the rock compass ended up pointing, then I could calculate exactly how much it had spun on the cable. Of course, I would be fooled if it had swung around more than one turn, but for less than one turn the answer would be clear.

A compass in the rocks. The presence of minute crystals of magnetite makes many volcanic rocks magnetic, but they have this magnetism only when they are cold.

This same technique can be used to determine whether blocks of rock have spun around or rotated within the Earth after they have retained their memory of the Earth's magnetic field, a rotation that is a distinct possibility wherever there are major movements in the crust.

It is not only once-molten igneous rocks that can acquire a magnetism pointing to the Earth's magnetic poles. Sand and silt also contain magnetic grains washed out of the bedrock by the action of rivers. As these grains are deposited in sedimentary layers farther downstream or on the floor of the ocean, some of the magnetic grains are free to move slightly in the watery pore spaces of the sediment. Each magnetic grain behaves like a floating compass needle and slowly rotates to align itself with the direction of the Earth's magnetic field. Eventually, as the sediment is buried and compressed, new minerals crystallize, filling the free pore spaces and preventing the magnetic grains from moving again. In this way, the sedimentary layers become, just like the cooling magma, a rock compass that can be used to detect any rotation of the rock body.

Drilling Holes

This is where my rock drill comes in. I had had it built in the department workshops in Oxford, copying the design of a visiting Norwegian geologist—Trond Torsvik— who had cleverly adapted a small chain saw engine. I showed Derek, our chief technician, the

Norwegian's drill. "Could you make something like this?" Derek looked at the strange steel contraption mounted on the engine, tilting his head from side to side as he tried to get the measure of it. "I don't see why not." And he did. This beautifully made piece of machinery, turned on the department's metal lathes, served me well in the rugged conditions of the Andes for many years. The actual drill is a rotating hollow steel cylinder, about two centimeters in diameter. Industrial diamonds, bonded to the end of the cylinder, give it a cutting edge; at high speed this edge can drill into almost any rocky surface, though the friction would make the steel cylinder dangerously hot if it were not continually cooled with water pumped down the barrel. I wanted to drill out a core of rock about ten centimeters long; this would be analyzed back at the laboratory in Oxford.

Drilling the volcanic lavas in the subzero temperatures of the high Western Cordillera, north of Ollague, turned out to be more difficult than I had anticipated. We had set off early in the morning, following the railway line north, on a rough track through a wide valley between volcanoes. The vast, dark peaks of Auchinquilche, rising over 6,000 meters, loomed above us, and though the rays of the sun were casting a golden glow on the volcano's flanks, we were still in deep shadow at the valley bottom. Where the track crossed the bed of a dried up creek, fresh-looking slabs of lava seemed to invite sampling. I shivered in the chilled air as I unpacked the drill. The water container rattled, full of a mushy ice, and the moisture on my fingers immediately froze when I touched the lava. I winced with pain, losing some skin as I tried to take my hand away again.

Surprisingly, the drill started the first time, with only one pull of the starter cord. The valley soon echoed with the whine of the small engine revving, sounding like a demented hornet. The pitch dropped to a growl as the rotating drill bit made contact with the hard rock, grinding out a circular groove. Leonore, who was acting as my assistant, started operating the water pump, vigorously moving the plunger up and down. A stream of runny mud began to squirt out of the drill hole, angling straight into my face and splattering me with rock fragments. As I pushed down hard, the drill inched forward, slowly cutting out a perfect cylinder of rock. Leonore must have decided to have a rest, because the muddy fountain tailed off. I screamed above the noise of the drill: "More water, more water." We soon emptied our first water container. I was very conscious of the fact that when the water runs out there can be no more drilling. In this arid and freezing environment, getting water was going to be a problem.

I drilled a whole succession of holes, working my way along the outcrop in the bank of the creek. It was only later that I discovered that despite all the heat generated by the friction of the drill bit, the drill water was soon freezing in the drill hole, bonding the rock core to the surrounding rock. After fiddling around with a screwdriver and hammer, cutting my freezing fingers, I realized that I had a big problem. In the end, the only way to get the cores out was by pouring boiling water from a thermos flask into the drill holes immediately after drilling. I have been told that scientists working in Antarctica overcome this problem by using paraffin, rather than water, as a coolant. However, the thought of paraffin squirting into my face has never inspired me to adopt this solution in Bolivia.

Eventually I freed enough cores. But before I could finally remove them from the hole, I had to carefully record their orientation. This was the whole point of the exercise, because I was using the cores as a sort of compass to record any change in the rock's orientation since the lava first cooled. I did this in much the same way that an engineer might work out the direction of an underground tunnel, using a compass to measure the bearing of the tunnel, and a tiltmeter to determine its slope. As each core was measured and safely put away, I felt more and more able to relax. When we were packed up again and ready to move on, we celebrated our first sampling locality in this remote region with tea and chocolate. It was nearly eleven o'clock anyway—the time for morning tea in the department back in Oxford.

We were very much alone when we were working among the volcanoes near Ollague—I don't remember seeing anyone all day. But in the Eastern Cordillera of Bolivia it is a different story. Here, there are plenty of people living on the rugged mountainsides, and drilling always seems to attract a crowd. I have often been surrounded by the local neighborhood—usually young children and old men—who show amazement at the sight of a group of foreigners apparently pushing a piece of metal into solid rock, creating a fountain of mud and water in the process. Sometimes, before the core has been measured and recorded, a child will dash forward, then nimbly snatch the core from its hole and run away with it. Once, a group of children on their way back from school stopped to watch. They were carrying pan-pipes and began to play an impromptu concert around us, while the youngest made repeated attempts to steal the freshly cut cylinders of rock out of my drill holes. I felt a bit like a parent at a rowdy children's birthday party, and it became increasingly difficult to concentrate on what I was doing with so much unpredictable activity going on.

Rock Cookery

Back in Oxford, I measured the magnetism of the small cores of rock, so laboriously drilled and orientated in the field. This too is a tedious and lengthy operation, and I have spent many long nights, when the only noise in the building is the hum of the laboratory instruments, sitting in front of a computer screen. The measurements are made with what is called a cryogenic magnetometer, an instrument that has its detectors bathed in liquid helium, at a temperature only a few degrees above absolute zero (−273.16°C). Keeping the detectors at this temperature makes them extremely sensitive, capable of measuring minute magnetic fields. For each specimen, the strength of the magnetic field must be measured in three different directions. Multiply this by the number of specimens, and, as we shall see, the need to repeat the analyses many times, and one ends up with thousands of measurements. Putting this information together, though, reveals which way the internal rock magnetic compass is pointing.

Geologists sometimes liken the study of rock magnetism to cookery. The people who make the measurements are called paleomagnetists (those who study ancient magnetism), or, less kindly, paleomagicians. The cookery or magic stems from the practice of heating up the rock sample in an oven to a specified temperature and then repeating the magnetic measurement. Unfortunately for an overworked scientist, it is necessary to repeat the measurement as many as ten times for a whole range of temperatures up to nearly 700°C. The reason for this is that rocks acquire magnetism not only when they first form, but also at later stages in their history. For example, when a rock is exposed at the Earth's surface, the minerals in it react with the atmosphere—this is the process of weathering. These chemical reactions can destroy old magnetic minerals and create new ones; enough to confuse the initial rock compass.

Heating the rock is often enough to break down the new magnetic minerals, removing the complications of later or secondary magnetism. The temperature when the magnetism is lost is often quite low, sometimes much less than 400°C; while that for the original magnetization, acquired when the rock formed, may be nearer the Curie temperature for pure magnetite at 580°C. In this case, if the rock is heated up to somewhere between 400°C and 580°C, it will lose any secondary magnetism but retain its original magnetism. This is often referred to as magnetic "cleaning." Another way to "clean" the sample is to subject it to varying magnetic fields, rather like erasing a tape in a tape recorder, and so remove the weaker, secondary magnetic field directions in the rock. All this cleaning has provoked some geologists to remark that the cookery lies not only in baking the samples, but also in cooking the results!

A Drunken Pole

A central assumption in the technique of using magnetism to detect the rotation of rocks is that this magnetism originally pointed to the magnetic North Pole. Unfortunately, the magnetic North Pole is not fixed itself but tends to move around. This phenomenon, known as secular magnetic variation, gives rise to the well-known deviation of magnetic north from true north, annotated on the edge of a detailed topographic map. It is a consequence of the fact that the Earth's magnetic field is inherently unstable. Geophysicists believe that the field originates deep inside the Earth, in the liquid metallic core. Here, molten iron swirls around in a complicated and fluctuating pattern, acting as a dynamo that generates a continually varying magnetic field.

The spin of the Earth stabilizes the dynamo to a certain extent so that the magnetic North Pole roughly lines up with the geographic North Pole. Over time, though, the magnetic pole seems to embark on an irregular walk, like that of a drunken man around the proverbial lamppost, as it were, of the geographic pole. At any instant, the magnetic North Pole can be as much as several thousand kilometers from the geographic pole. For example, as I write this book in 2002, the magnetic North Pole lies in the Canadian Arctic on Ellef Ringnes Island at a latitude of 78°N, which is about 1,300 kilometers from the North Pole. Since 1831, when it was first accurately located by the arctic explorer James Ross, it has moved on average about fifteen kilometers each year. Detailed measurements of the position of the magnetic North Pole over much longer periods, using the magnetism in rocks, has shown that it completes its drunken perambulation around the geographic North Pole in a few thousand years.

Secular variation greatly complicates the way studies of rock magnetism are carried out. For example, the direction of the magnetic compass from a single sample of a volcanic lava flow is difficult to interpret. The lava may have erupted over a period of a few days and so the internal rock compass, imprinted when the lava cooled, will point to the magnetic pole for that geological instant in time. The volcano may then lie dormant for a few hundred or even thousand years. However, the magnetic North Pole would continue to take its drunken walk of secular variation. Thus, when the volcano bursts into life again, spewing out new flows of lava, the new rock compass will point to a new position of the magnetic North Pole. During this geological short history of volcanic activity, there will almost certainly be no substantial movement or rotation of the volcano. Yet, the magnetic measurements, if taken at face value ignoring secular variation, would suggest otherwise.

The trick, when dealing with secular variation, is to look at a whole series of snapshots of the magnetic North Pole on its drunken walk, averaging the measurements to estimate the position of the "central lamppost" of the geographic pole. This is simply done by sampling a large number of different volcanic eruptions or numerous layers of sediment from a sequence of sedimentary rocks, formed over a period of several thousand years, and enough time for the magnetic north pole to complete its drunken perambulation. This strategy is easier to describe than actually carry out. For example, the only way to do this in the foothills of the Andes, where thick sequences of red-beds are best exposed in the banks of wide, sandy rivers, is to stand chest deep in the water and drill numerous holes in a whole succession of rock layers exposed along the riverbank. The cool water is a luxurious contrast to the hot and humid atmosphere, but handling the drilling in equipment in these conditions is extremely tricky. Once, a special tool, which I used to disconnect the drill barrel from the engine mounting, fell into the muddy river water. I spent twenty anxious minutes probing the sandy river bed with my feet. Eventually, to my relief, I felt the sharp metal edge of the tool against my toe and ducked underwater to pick it up.

I was principally interested in the rotations to do with the pushing up of the Andes, and so it was important to separate out the large-scale plate motion of South America as it has drifted over the face of the Earth. This drift can be studied by looking at the magnetic properties of rocks exposed in South America, outside the Andes, averaging out the effects of secular magnetic variation. This way, one can construct what is called an "apparent polar wander path" for South America over the last few hundred million years. This also seems to show that the North Pole has moved around, but, because of the way that we have now averaged the measurements, the movement cannot be the result of the magnetic pole's drunken walk. In reality, it is South America and not the North Pole that has moved. But for rocks of the same age, the difference in the apparent position of the North Pole, when viewed from either the South American tectonic plate or rocks in the Andes, must be a consequence of relative motion between the two regions—in this case, the result of local rotation of rocks in the Andes.

The Athlete Crouches

Returning to my analogy of the athlete at the beginning of this chapter, I had planned to collect rock samples for magnetic analysis from the leg, above and below the knee of the Bolivian orocline. I also wanted to analyze rocks with a wide range of ages, laid down or erupted at various stages during the growth of the Andean mountains in the last forty million years. If the athlete had originally been standing upright, but then had crouched down, bending the leg, I would expect to see this in terms of the realignment of the internal rock compasses. Huge tracks of the Andean mountains above the knee, in the thigh, would have pivoted round in an anticlockwise sense, or the mountains in the calf, below the knee, would have pivoted in a clockwise sense, or there would have been some combination of both these rotations. On the other hand, if the athlete had always been crouching, then the rock compasses in the thigh and calf should show no sign of bending but consistently point in the same direction and parallel to rock compasses outside the Andes.

When I started this work in 1990, there were very few published magnetic measurements of the rocks exposed in the Bolivian Andes. These were difficult to interpret, but there was a suggestion that the athlete had indeed crouched down from an upright position. To confirm this, I had to find more magnetic rocks. This turned out to be my biggest problem; it was only when I had taken the rocks back to Oxford and analyzed them that I discovered if there was any coherent magnetism at all. I have spent hours or days sampling the rocks, only to learn a few months later that I was completely wasting my time—either the magnetism turned out to be far too weak, even for the supersensitive cryogenic magnetometer, or it had been completely wiped out by later chemical reactions in the rock. This made me rather cynical in the field, and I tended to be overcome by pessimistic thoughts: "this will never work" or "there is no point drilling here, the rocks are far too altered." Pessimism is a bad trait in a field geologist because it discourages curiosity. There have been times when, despite my worst fears, the rocks were bristling with important magnetic information.

My first field season in 1990 turned out to be a big disappointment—in most cases I failed to find anything that could be used as a rock compass to detect rotations. Was it worth carrying on? The research was time consuming and expensive. Perhaps I should just cut my losses and try something else? Somehow, giving up wasn't really an option, and so I soldiered on. Gradually, as is often the case in research if you persevere, the whole endeavor started to look more promising. Over the years between 1990 and 1995, I managed to accumulate rock compasses from above and below the knee of the Bolivian orocline, and a pattern of bending of the mountains on a gigantic scale began to emerge. I was not alone in this work. In some ways, this was rather galling: my research was moving forward so slowly, like a tortoise, and I felt that I was being rapidly overtaken by other harelike scientists.

There were two in particular—Pierrick Roperch and Bruce MacFadden. Bruce MacFadden was a paleontologist who was using the magnetism in red-beds to characterize and date particular sequences of layers in which his fossils had been found—a technique known as magnetostratigraphy. He had started working in Bolivia in the 1980s, trying to track down the evolutionary origins of horses, and with each new season of fossil collecting he would often obtain—almost as a by-product, I suspected—some useful paleomagnetic data about rock rotations. Pierrick Roperch was a French geophysicist who ran a laboratory at the University of Rennes specializing in analyzing the magnetism in rocks. Whereas MacFadden had started his work before me, Roperch had begun his research in Bolivia a year or two later. But Pierrick soon turned into a prolific researcher, drilling literally thousands of cores all over Bolivia—I sometimes came across his drill holes when I was doing my own sampling. I would also bump into Pierrick at conferences, though I felt he was rather secretive about what he was up to. He probably felt the same about me.

The outcome of all this effort was very simple but at the same time enormously significant, and there was no other way of finding this out. North of the knee, in the thigh of the Bolivian Andes, the rock compasses in sediments laid down around sixty-five million years ago, before the birth of the Andes, were systematically deflected counterclockwise of north toward the eleven o'clock position—roughly thirty degrees. In contrast, the same rocks in the calf, farther south, were deflected clockwise toward the one o'clock position. These deflections, when compared with our reference rock compass in the lowlands of South America, suggested that virtually the entire bend in the knee had been the result of our Andean athlete crouching since then. Younger rocks show less of this crouching, catching the Andes in the act, so to speak, of bending. And it is not only the mountains in Bolivia that have swung around like this; paleomagnetic studies now show that the same thing has happened much farther north, in northern Peru and Ecuador, and to the south in Patagonia.

It is a staggering thought, when viewing the overall sweep of the coastline along the western margin of South America, that much of its sinuous course must be the result of the bending of the vast Andean mountain range, as it has risen up in the last few tens of millions of years. This is a graphic example of how movements in the Earth's crust can change the face of the planet. Yet it is also clear that the same bending has not gone on in the interior of the South American continent. So, how is it possible to bend one side of a continent without affecting the interior? The answer is quite simple. You have to create a mountain range like the Andes.

These mountains are highest and widest at the knee of our athlete. And the roots of the crust are deepest here too. The conclusion must be that the bend at the knee is the result of intense squeezing of the crust, and the amount of squeezing becomes less farther north and south. It is now easy to understand what has happened. We can imagine the continent of South America to be a giant slab of clay. The western side of this slab has been molded into a new shape, prodded and pushed, if you like, by a sculptor's invisible fingers. And the largest push of all occurred right in the knee of

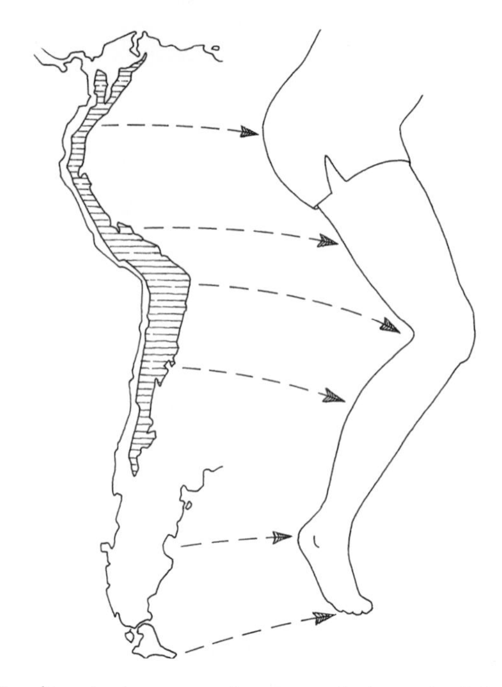

Bending the leg: squeezing the rocks has raised the Andes. But this has also remolded and twisted the western margin of South America, creating the distinctive bends in the coastline, including the knee of the Bolivian Andes.

the Bolivian Andes. This way, the mountains were raised and the coastline changed shape.

Out of the Doldrums

With this final view of the rise of the Bolivian Andes, we have arrived at a surprisingly full account, described in the preceding chapters, of the huge movements in the crust involved in their evolution, resulting in a change in the shape of the whole western side of South America. Putting all this research under our belts, now is the time to return to the history of ideas about mountain making—a story I started in an earlier chapter, before recounting my own first visit to Bolivia. By the beginning of the twentieth century, geologists had somehow lost their way, confused and unable to make sense of conflicting observations and theories about the mobility of the Earth's crust.

Alfred Wegener had suggested in 1915 that the continents have moved over the surface of the Earth, sometimes rifting apart, sometimes colliding. British geologist Arthur Holmes—a pioneer in measuring the age of the Earth—speculated in 1929 that these motions were driven by giant convection currents deep inside, related to the upwelling of hot rocks and the sinking of cold rocks. But all these ideas had been rejected out of hand by the scientific community; geophysicists were adamant that the interior of the planet was far too strong to flow like this. Yet the evidence for huge movements on the surface refused to go away. And so, for many decades, geological thinking had become like a ship, becalmed in doldrum seas unable to make headway in any direction. It was not until the late 1950s, at the height of the Cold War, that the American military gave it the means to set sail again. Though the story of how this led to the theory of plate tectonics—a theory that is crucial to explaining the movements I had observed in the Andes—has been told many times, it is worth retelling once again. To begin with, it involved both earthquakes and appropriately, given what I have already described in this chapter, the magnetism in rocks.

All at Sea

The signatories of the Nuclear Test Ban Treaty paid for the installation of a worldwide network of earthquake listening devices (seismometers), designed to detect unauthorized nuclear explosions. These seismometers were soon swamped by thousands of natural earthquakes. Detecting a manmade nuclear explosion was going to be like looking for a needle in a haystack. But studying the natural earthquakes was easy. It soon became clear that earthquakes in the oceans are largely restricted to long, narrow zones that snake across the Earth. Long sections of these zones follow a hidden secret of the oceans—the midocean ridge—a nearly continuous undersea mountain range that encircles the planet, rising 2 kilometers above the deep ocean floor. In the Atlantic Ocean, a deep, narrow groove or valley, only a few tens of kilometers wide, runs along the center of the midocean ridge. Numerous deep sea dives in miniature underwater submarines, designed to cope with the immense water pressures, have now shown that this valley is rather like an underwater version of the great rift valley of East Africa, full of active underwater volcanoes erupting lava. Black clouds of superheated water gushing out of deep cracks—black smokers—testify to the presence of very hot rocks not far beneath the sea floor.

The midocean ridge is where the sea floor is literally born. This extraordinary idea had been put forward by Arthur Holmes in the 1920s, but it was taken seriously only in 1963, when British geophysicists Fred Vine and Drumond Mathews published a short paper in the scientific journal *Nature*. Their work had been inspired by some remarkable discoveries about the Earth's magnetic field. To begin with, the United States Coast Guard had agreed to let geophysicists tow a new type of extremely sensitive device—a proton magnetometer—behind one of their ships while they were surveying the offshore parts of the western United States. To everybody's surprise, as the ship criss-crossed the ocean waters, the magnetometer revealed regular differences from place to place—referred to as anomalies—in the strength or intensity of the Earth's magnetic field. It was as though the ocean had been dyed with the black stripes of a zebra, or a bar code, corresponding to a series of thin, long zones in which the field was either stronger or weaker than normal. Soon, these stripes were found in all the world's oceans, and it became apparent that they always ran parallel to the midocean ridge.

Vine and Mathews had a hunch that the magnetic anomalies reflected variations in the magnetism of the lava on the sea floor. They knew the basic principles of rock magnetism—when a volcanic lava flow solidifies, it becomes slightly magnetic and behaves a bit like a bar magnet or compass pointing to the Earth's magnetic North Pole. They knew something else: the magnetic North Pole was extremely unstable and had even, on occasion, completely flipped or reversed, so that a compass needle would point south rather than north. The last time that this had happened was about three quarters of a million years ago, but there had been many other flips before this. Vine and Mathews's great insight was to put all these ideas together. If the lava cooled when the Earth's magnetic field had undergone one of its flips or reversals, so that magnetic north became south, the rock would also acquire a magnetism in the opposite direction. They calculated how reversals like this in the magnetism of rocks on the sea floor would affect the magnetic field at the sea surface—there was an almost perfect match with the ocean magnetic surveys. The sea floor seemed to be behaving like a tape recorder, picking up the flips in the Earth's magnetic field as a series of thin stripes. To Vine and Mathews, the only way this could happen was if the sea floor was being continually created by volcanic eruptions at the crest of the midocean ridge, imprinted with the direction of the magnetic pole before cooling and steadily moving away like a giant conveyor belt.

This revolutionary vision of a mobile sea floor—sea floor spreading, as it came to be called—opened up a whole new way of looking at the Earth's surface and was a huge leap forward on the road to plate tectonics. But first, geologists had to answer another question: what happened to the sea floor as it moved away from the midocean ridge? If the size of the Earth remained constant, as most geologists thought, then there must be places where the sea floor was being destroyed rather than created. What these zones of destruction might be was dramatically revealed when geologists started examining another extraordinary feature of the planet—the Ring of Fire.

The Ring of Fire

The Ring of Fire is a gigantic necklace of volcanoes that extends virtually right around the edge of the Pacific Ocean. It extends north from New Zealand to Fiji, the Solomon Islands, and the Philippines, before swinging east, through Japan and along the Kurile and Aleutian Islands to the southern coast of Alaska. From here it turns south along the west coast of North America, and then on south to South America, following the Andes all the way down to southern Chile. Throughout the Ring of Fire there is a remarkable association among earthquakes, volcanoes, and a giant groove or trench in the sea floor. A good place to start investigating these features is on the western margin of South America.

The Spanish conquistadors were the first Europeans to feel the earthquakes that periodically rock the Andes. By the beginning of the twentieth century, the Jesuit mission in South America had set up seismological observatories to monitor them. One of these observatories, the Observatorio San Calixto, is still operating in La Paz, Bolivia. But it is no longer necessary to be based in South America to study the earthquakes. Since the 1950s, when detailed information about the earthquakes first started to become available, thousands of earthquakes have been recorded. It has become clear that they tend to cluster. Hugo Benioff, an American seismologist, was the first to have a hint of this. By 1954 he had located enough earthquakes to realize that they mostly lay in an inclined, slablike zone, only a few tens of kilometers thick, which extended from near the Earth's surface, offshore, down to hundreds of kilometers beneath South America. The remaining earthquakes occurred at relatively shallow depths beneath the mountains.

The Andes were not the only place where Benioff detected dipping zones of earthquakes, extending deep into the Earth. In fact, these zones extend virtually around the entire margin of the Pacific ocean. Geologists soon dubbed them Benioff zones. The shallow portions of Benioff zones, where the zone of earthquakes begins to plunge into the Earth, was found to coincide with a remarkable feature of the sea floor—deep ocean trenches, first discovered on the great scientific voyages of ocean exploration in the 1870s. These trenches form a deep groove, in a series of graceful arcs, lying roughly a hundred kilometers offshore, and, again, running around most of the Pacific. Off the Mariana Islands in the western Pacific, the bottom of the trench reaches a staggering 11 kilometers below sea level. However, along the western margin of South America, the ocean trench reaches a more usual depth of about 7 kilometers. In some ways, the height of the Andes should really be measured from the bottom of this trench, making many of the mountains well over 13 kilometers high, and forming the largest continuous vertical relief on the planet.

Subduction Zones

The full significance of the ocean trenches and Benioff earthquake zones became clear only when the 1964 magnitude 9-plus earthquake struck the coastal parts of southern Alaska. Immediately afterward, George Plafker, a United States Geological Survey geologist, had the good sense to go and look at the changes in land levels that had taken place around Prince William Sound and Kodiak Island. In 1996 I spent several weeks with George, returning to this region. An extremely tall and distinguished looking man, he was pleased, I think, to have the opportunity of reliving the excitement of those days in the aftermath of the disaster with somebody as enthusiastic as himself about its impact on the landscape. Much of the evidence was still there to see, thirty-two years later. We found the remains of great banks of barnacles, tenaciously clinging to the rocks, which had risen up out of the water during the earthquake. And elsewhere, where coastal regions had been inundated with sea water and icebergs, the ruins of abandoned cabins, with wallpaper hanging loosely from the walls and an odd child's shoe on the floor, stood as reminders of the disaster. By marking on a map how much any particular bit of coastline had gone up or down, George had been able to construct an accurate picture of the profound changes to the land. It was clear that a wide and long coastal region, about the size of Britain, had been uplifted up to 12 meters—more than the height of a four-story building—while farther inland, another similar sized region had subsided up to 2 meters. All this in the few seconds of the earthquake!

The most puzzling feature of George's work was that he had failed to find any evidence for a large break or fault in the Earth's crust, despite the expectations of most geologists that earthquakes

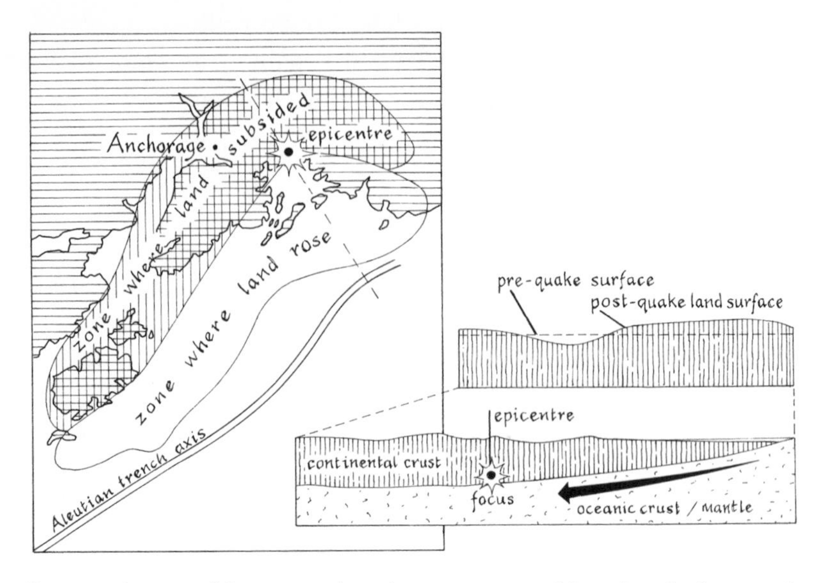

In 1964 the second largest earthquake ever measured by seismologists struck the south coast of Alaska, causing large vertical shifts in the landscape. The pattern of uplift and subsidence can be explained only by slip on a gigantic and gently inclined megathrust beneath southern Alaska, reaching the surface in the deep oceanic trench offshore.

were the result of sudden slip on a fault. George told me that he had long puzzled about this, or, as he actually put it: "It really bothered me." The realization finally dawned that the reason he could not find such a fault was that it was hidden, underlying this part of Alaska as a giant inclined plane and reaching the surface only out to sea in the deep ocean trench. Such a huge fault had to be called a megathrust! And movement on it was not a one-time event. There were scars in the landscape that testified to many previous earthquakes, with each earthquake occurring at intervals of eight hundred years or so. During the long period between earthquakes, forces in the Earth must build up as the Alaskan crust is squeezed like a spring. George had realized that this would push up the rocks above the megathrust into an elongate bulge. Eventually, the crust would reach breaking point, suddenly slipping up the fault toward the Pacific Ocean and raising the coastal parts of Alaska, but bringing down the bulge farther inland. Today, the Alaskan crust is being slowly squeezed again, in preparation for the next earthquake.

Inspired by his discoveries in Alaska, George had started wondering if other large earthquakes in the Ring of Fire could also be due to slip on megathrusts. He managed to persuade his boss at the U.S. Geological Survey to release him from his geological duties, so that he could he spend a year in southern Chile, near Concepcion, studying the region that had been devastated in the huge 1960 earthquake—the largest earthquake ever recorded by seismologists. This was the same stretch of coastline, extending for nearly a thousand kilometers, that had been shaken in the earthquake during Darwin's visit in 1835. And, indeed, the land-level changes were very similar to those George had observed in Alaska after the 1964 earthquake, with uplift of the coast and subsidence farther inland, suggesting slip on another megathrust. As all motion is relative, George realized that it was easier to think about movement on the megathrusts in terms of the underlying Pacific Ocean floor sliding beneath the Ring of Fire and sinking into the mantle. Seismologists have now monitored many other, smaller earthquakes in the Ring of Fire. They can use the pattern of earthquake-triggered vibrations to determine exactly how the fault has slipped, revealing the same sinking motion along megathrusts around the edge of the Pacific Ocean.

By the late 1960s, geophysicists studying the sea floor had come to realize that it was not just the ocean crust, but a slab of the outer part of the Earth, about a hundred kilometers thick, that was sinking. The ultimate fate of the sea floor, created along the midocean ridge, now became clear. As it approaches the edge of the continent, it bends down in the ocean trench and slides beneath the megathrust, beginning its journey into the Earth's interior. This is a subduction zone: the ocean floor is being subducted. And it was obvious why the ocean trenches lie in arcs: these arcs are an inevitable consequence of the bending of the surface of a spherical Earth—one only has to push in the rubbery surface of an inflated

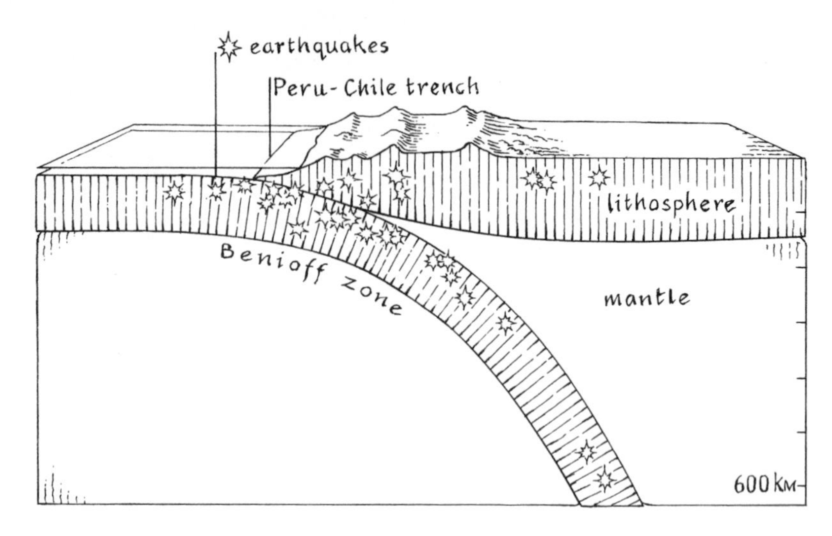

The western margin of South America is rife with earthquakes. These occur mainly in an inclined, slablike zone that plunges beneath the Andes, called a Benioff zone after the seismologist who discovered it. The earthquakes are tracking the ocean plate as it sinks into the Earth, bending down in the ocean trench and sliding under the Andean ranges.

balloon to see that every depression has circular edges. The ocean floor that is being subducted beneath the Andes is part of what geologists call the Nazca plate; for most of the rest of the Pacific margin there is subduction of the Pacific plate. The Benioff zone is merely a trail of small earthquakes in the sinking slab, triggered as it is subjected to the forces inside the Earth. This zone traces the slab down to a depth of 700 kilometers in the Earth's mantle, revealing the deep structure of the subduction zone. At greater depths, the rocks have become too weak, as they slowly heat up inside the Earth, to give rise to earthquakes.

Two Steps Forward, One Step Back

Put sea floor spreading together with subduction zones and one has essentially Arthur Holmes's idea of an Earth made up of giant convection cells, with their upwelling and downwelling limbs of hot and cold rock. The midocean ridges overlie the upwelling of hot rock from deep in the Earth. At the surface, this hot rock continually melts to fuel the volcanic eruptions that create the sea floor with its magnetic stripes. As this happens, the sea floor moves steadily away from the midocean ridge, eventually sinking back into the Earth's interior at subduction zones. Holmes, in his original 1929 publication, illustrated this pattern of convection with a sideways view inside the Earth.

Now, geologists, armed with the global pattern of earthquakes, realized that the convection cells, if viewed from above, would divide the surface of the Earth into a number of rigid, moving regions. And so the theory of plate tectonics was born. This simply states that the surface of the Earth is made up of a mosaic of rigid plates that are in constant relative motion. As most of the boundaries to these plates lie on the sea floor—midocean ridges, subduction zones, and connecting faults called transforms—it is no wonder that land geologists had failed to come up with the theory. The plates move, from a human perspective, very slowly, at a speed amounting to not more than several centimeters each year—about the rate your fingernails grow. But over geological time the displacement builds up, so that during the long history of the Earth the tectonic plates have had time enough to wander all over its surface. The continents form part of the plates and move with them, sometimes drifting apart, opening up new oceans, and sometimes moving together and colliding, closing old oceans. In this way, the motions of the plates reshape the face of the Earth.

The theory of plate tectonics turns out to be the essential starting point for any understanding of the origin of mountains. But, as we shall see in the rest of this book, it is not the whole story, and, indeed, for some parts of the Earth, we will need to abandon this theory altogether. We can see the scale of the problem when we look at the places on Earth where two plates approach each other, either as colliding continents or in subduction zones. Mountains do not appear to be an inevitable consequence of this "coming together," and each mountain range is unique, varying widely not only in its overall height and extent, but also in its geological history.

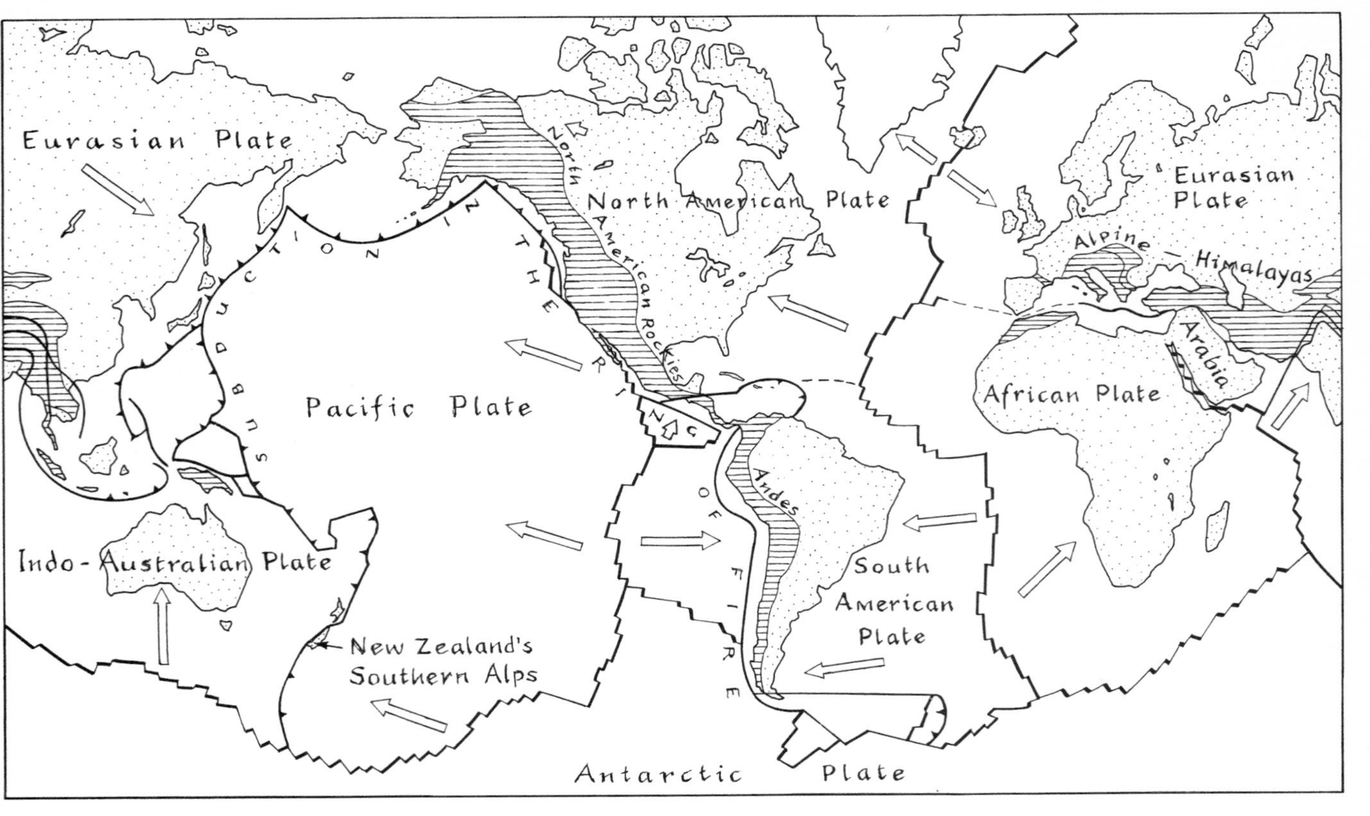

The surface of the Earth is divided up into a mosaic of moving tectonic plates. The continents are part of the plates and move with them. The planet's great mountain ranges occur where the plates (either continent or oceanic) are converging—the oceanic Nazca plate is sliding beneath the Andes along the western edge of the South American continent.

Along the Alaskan subduction zone, the mountains are really no more than deeply eroded stumps, and there is no evidence that they are being pushed up today. In New Zealand, where the earthquakes would suggest more active squeezing of the crust, the ranges are still pretty narrow and low, only a few tens of kilometers wide with peaks that generally do not rise above 2,000 meters. But, in marked contrast, the Andes form a vast mountainous tract, in places 700 kilometers wide and rising well over 4,000 meters, that follows the subduction zone on the western margin of South America for thousands of kilometers. In Central Asia, where there is a more complicated arrangement of subduction zones and convergence of continents, the mountains of Tibet and the Himalayas extend over a region that is several thousand kilometers in both length and breadth, rising up to nearly 10 kilometers above sea level. Clearly, we need more than just the theory of plate tectonics to understand this myriad of variation among mountains.

The great chains of volcanoes, which are such a prominent part of the mountains in the Ring of Fire, are also unexplained by plate tectonics. Yet the volcanoes are the feature of the Ring of Fire that gives it its name in the first place—a ring of notoriously violent and explosive volcanic eruptions. In South America, the volcanoes form huge conical mountains, reaching altitudes over 6,000 meters above sea level in southern Peru, northern Chile, and Bolivia. They are liable to erupt at any time, producing a characteristic dark volcanic lava called andesite. In fact, andesite is the typical volcanic rock throughout the Ring of Fire, but its origin is still debated. And it has been unclear to many geologists what role these rocks, either when they erupt at the surface or when they cool slowly at depth, have played in building up the mountains.

To me, it is this very uncertainty that makes the Andes so fascinating. I remember thinking, when I first came to study them, that the greater the challenge, the more I would be able to achieve in my own scientific career—a world in which everything had been explained by plate tectonics seemed boring and dull. The flush of excitement that plate tectonics had brought really belonged to a previous generation of geologists. Nevertheless, I could never escape from the fact that these mountains stood on the boundary between two great tectonic plates that were relentlessly moving toward each other — an idea that many would have thought ludicrous before the theory of plate tectonics came along.

✣ ✣ ✣ ✣ ✣

More Rough Roads

Our old and faithful Toyota Land Cruiser took an immense pounding in the field, especially when it was heavily laden with people, gasoline, rocks, and equipment. When we eventually arrived at the safe haven of a major town, I seemed to spend most of my time with mechanics, peering at the underside of the vehicle and trying to deal with another mechanical problem. The suspension was always a problem: this was the front line in the battle of the roads. The backbone of the suspension system was a set of leaf springs — a stack of arcuate metal bars. These flex every time the vehicle goes over a bump. However, if they flex too rapidly, they crack and must be replaced. The local mechanics have given each metal bar in the leaf spring a number. The biggest bar is the one that actually links the axle to the chassis: this is called the *primera hoja* (first spring). It was not until we had broken our first primera that I truly felt initiated into the Bolivian road system. In effect, one end of the back axle was no longer attached to the vehicle. In our case, it was only the exhaust pipe that was holding the axle on, and one of the back wheels was jammed right up against the wheel arch. But because the engine was so powerful, we managed to drive for a considerable distance without realizing that anything was wrong.

On another occasion, we were heading across the desolate Altiplano toward the border with Chile. At that time Bolivia was virtually at war with Chile over a disputed international border. The Bolivian government claimed that a group of Chileans had moved some of the border markers. To prevent further movement, mine fields had been laid along the border zone, and both governments had deliberately left the main roads close to the border in a bad state. In fact, there was no road as such leading from Bolivia to Chile, but a maze of wheel marks where truck drivers had blazed a trail around the mine fields and through the barren and sandy volcanic landscape. Yet almost all the imported goods in Bolivia had to enter the country this way, trucked from the Chilean ports of Antofagasta or Iquique. I remember seeing a huge open truck, loaded high with pink toilets, struggling through the landscape. Its wheels were deep in the volcanic ash as it swayed from side to side, like a ship at sea, traversing the rough ground.

At several points on this route it is necessary to cross deep rivers. The big trucks can easily drive across them, but we decided that one of the rivers was just too deep for us. Then, out of nowhere, a local man mysteriously appeared on a bicycle and guided us to a shallower ford where we could cross. By now the underneath of the vehicle had been thoroughly immersed in freezing cold water. It was getting late; time to set up camp for the night. We decided not to risk getting stuck in the sand or driving over a land mine, and so we parked the vehicle on the track. There were so many of these tracks, which, in any case, were not much more than two wheel marks, that we thought nothing of it. We were soon snuggled up in our sleeping bags as the outside air temperature plummeted to well below zero. About two o'clock in the morning we were awoken by the roaring sound of a diesel engine close by. I looked out of the tent to see a huge truck towering above us. We decided to ignore it. Eventually, it went around us, sinking deep in the sand, and we went back to sleep.

At about four in the morning we were awoken again by the sound of another truck engine. We were evidently camping on a popular set of wheel marks. This time the driver refused to go around. The passengers—campesinos on their way to market—swarmed out of the back of the truck and insisted that we move. I had no choice but to comply. However, as I slowly let the clutch out, the engine shuddered and stalled. I was surprised but tried again with a more powerful low-ratio gear. The same thing happened: the vehicle would not budge. By this time the truck driver had lost patience, had tried to go around us, and was soon stuck deep in the sand. The passengers quickly collected armfuls of the local tholla scrub and piled it under the wheels. With a grinding of gears and revving of the engine, the truck finally freed itself and made its way back to the very slightly harder sand on the other side of our camp, disappearing into the night.

The next morning I discovered why our vehicle would not move. The brake drums, soaked in the previous day's river crossing, had frozen solid in the night. As I pondered what to do, a local man emerged, yet again on a bicycle, following a narrow path through the desert. He passed close to our car and stopped to have a look. I explained that we were stuck because the brakes had frozen solid. He accepted this as though frozen brakes were quite normal and commonplace and headed off on his journey. I watched him dwindle into the distance. My first idea was to pour hot water over the brakes. Despite heating endless pots of water on our cooker and pouring them over the brakes, they remained as firmly frozen as ever. I then noticed that the handbrake cable was still taut, even though the handbrake was off. It was clear that the handbrake cable itself had frozen to the car. I tried disconnecting the whole handbrake system from the back wheels. Revving the engine and using the most powerful gear, I dropped the clutch. There was a loud crack, like a gun going off, as the ice finally broke and the car lurched forward. We were free at last. But I had learned an important lesson: never ford a deep river in the late afternoon if you are going to camp out for the night.

PART THREE

✣

✣ CHAPTER EIGHT ✣

A Sort of Fudge Cake

Fudge, ***n.*** Soft grained toffee-like sweetmeat made with milk, sugar, butter etc. (*OED*)

DUST EVERYWHERE. In my eyes, in my nose. I walked through the fine grey powder kicking up clouds with each step, using my handkerchief, wrapped round my face, as a mask. Around me, jumbled piles were all that remained of houses. It was sad to recall my last visit to Aiquile, a few years before, when it was a living town, boasting, with its stuccoed central plaza, its status as a market center in the Eastern Cordillera of Bolivia. Admittedly, it had been on the decline; the railway line lay abandoned and rusting, but there were always rumors that a new company would make the trains arrive again. Now these rumors seemed like a cruel joke, with Aiquile deserted and its survivors squatting in a field of blue tents, hastily provided by international relief organizations. I clambered over yet another heap of mud bricks and came face to face with an old man. "You people," he shouted. "You only come to stare, then you go away again. We have to carry on living here." His words made me feel ashamed—he was right. I had come here only to stare. I wanted to see for myself the destructive power of an earthquake. And this one, which had struck Aiquile in the middle of the night, only a few weeks ago, had been the deadliest for nearly a century, killing over a hundred people when the roofs came crashing down on top of them as they lay in their beds.

I rounded a corner and entered a street where a few buildings were still standing with a second story. Reinforced concrete, I thought. Most of the houses were adobe brick, held together with clay and straw. When the ground shook, these crumbled into dust. But just a bit of strengthening with concrete, and the walls of the house stayed up, though the wood and straw roof fell in. I wondered what the new Aiquile would be like, the town that would inevitably rise up from this chaos. Would anybody advise the locals to use concrete and cement, or would they just revert back to the age-old methods? Nobody had any money. So, without help, they really had no choice; mud houses were all they could afford. I came across a section of plastered wall, still upright. It had a political slogan daubed across it, ending abruptly like the edge of a torn page where the rest of the wall had collapsed: "Revolución es libertad." Two perfect cracks ran through the wall, intersecting in a cross. The slogan had been displaced a few millimeters across the cracks, producing an offset through the letter "R" of *Revolución.* I had, at last, come face to face with the natural phenomenon that had triggered all this destruction—sudden movement on a fault line in the Earth's crust.

I never found the main fault line that had ruptured during the Aiquile earthquake, despite looking hard for it in that rugged landscape. Perhaps it was buried too deeply and never broke the surface? But the twisted and broken rock layers that formed the bedrock of these mountains screamed out at me that the Aiquile earthquake was nothing. There had been many earthquakes before in this region, and these had been far more powerful, with enough destructive force to destroy tens of Aiquiles in a few seconds. I had seen the long-term effects on the landscape of earthquakes like these during my time in New Zealand in the mid 1980s, where I had spent three and a half years studying the active fault lines that sliced up this small fragment of land on the far side of the Pacific Ocean from South America. I had been taught by an extraordinary New Zealander, Harold Wellman, how to read the rich language of faults and earthquakes. This experience had whetted my appetite for the sort of geological research I was now carrying out in our Andean project. And I was aware that despite all our work, we still had a deeper scientific question to answer. We had shown that the mountains had been mainly squeezed up along faults, but why had this created something with the particular shape and heights of the Andes? As it is always best, when trying to answer difficult scientific questions, to go back to the beginning, let me start in New Zealand.

Harold Wellman

It still amazes me how lucky I was, not only to meet, but to spend a considerable amount of time together with Harold Wellman while he was still fit enough to go into the field—he died in 1999. He had a profound effect on me during an early stage of my research career. Harold was born in Somerset, England, in 1909 and came out to New Zealand with his parents in the late 1920s. Although he never returned to England again for any length of time, he still had a slight Somerset burr in his voice when I first met him in the mid 1980s. A tall man—this seems to be a characteristic of great field geologists (unfortunately, I am fairly short)—with the weathered face of somebody who has spent much of his life outdoors, he could be very rough and abrasive, though there was just enough gentleness underneath to convince me that he actually liked me! During the Great Depression he took part in one of the last gold rushes in New Zealand on the beaches of the west coast of South Island, where small grains of gold were to be found in the beach sands. Harold once summed up his gold-mining days to me by remarking that he earned about a New Zealand dollar a day (roughly equivalent to one U.S. dollar at the time he was speaking to me), enough to live on, but nobody was going to get rich very fast this way.

By now New Zealand was fighting in the Second World War, and Harold had joined the New Zealand Geological Survey as a field geologist. He was sent back to the west coast of South Island, revisiting the scenes of his early gold-prospecting experiences, on a mission to find mica in the mountains of the Southern Alps. This was not an easy task because the landscape is cloaked in dense vegetation, watered by over eight meters of rain each year. Numerous bouldery creeks carry much of this rain back through the forest to the sea. Harold Wellman and his colleague, Dick Willet, had no option but to scramble up and down these creeks, looking for traces of mica in the river gravels. Their search was to lead to a discovery of something far more important than mica—a discovery that would change the way geologists thought about movements of the Earth's crust.

The western range front of the Southern Alps is remarkably straight and abrupt. Harold noticed a pattern in the types of boulders strewn over the creek beds. Where the creek was still in the mountains, the boulders were made up almost exclusively of a distinctive metamorphic rock called schist. This was hardly surprising as the bedrock of these mountains is made up of this very schist. As Harold proceeded downstream, he found a new type of boulder in the creek bed—granite. In places, knobs or small hills of granite poked up out of the bush on the coastal plains. The important discovery was made when the first occurrence of granite boulders in each creek bed was marked on a map. It followed a nearly straight line extending along the range front of the Southern Alps. Harold realized that the boulders in the creek were telling him about a change in the bedrock. In other words, the edge of the Southern Alps was not just an abrupt feature of topography, it also marked a boundary between two types of rock.

In a few creeks it was actually possible to see the junction in the bedrock between the granite and schist. Here, in the river bank, a greenish clay material outcropped between the granite and schist in a zone a few meters wide. Wherever the junction between the schist and granite was exposed at the foot of the Southern Alps, the same greenish clay band was found. To a geologist, a sharp, straight boundary between two rock types, with a thin seam of clay in between, can mean only one thing: a fracture or fault—Harold Wellman called it the Alpine Fault as it ran all along the foot of the Southern Alps. The scale of the Alpine Fault was immense, extending for over five hundred kilometers right up the west coast of the South Island. Of even greater interest was the movement that had occurred on this fault. Clearly, as the granite and schist had slid passed each other, the rocks had been pulverized into a greenish clay. Harold reasoned that this movement must have also pushed up the Southern Alps.

In 1942 Harold published his discovery of the Alpine Fault. By then he had realized that there had been colossal horizontal movement along the fault, on a scale that at the time most geologists thought was impossible. Miners had known for some time about a distinctive band of volcanic rocks at the southern end of the South Island, among the schist on the eastern side of the Alpine Fault. This band could be traced westward toward the Alpine Fault. Harold had himself discovered an identical band of rocks on the western side of the fault, at the northern end of the South Island. If the rocks were slid back horizontally about 480 kilometers, then the two bands matched exactly. This showed that the Alpine Fault was principally what geologists call a strike-slip fault, and movement along it had divided New Zealand in two, displacing sideways once continuous bands of rock nearly five hundred kilometers. However, there was also a vertical component of motion—this, together with movement on other faults, known as reverse or thrust faults, had pushed up the Southern Alps.

A Landscape on the Move

Most geologists outside New Zealand took little interest in Harold Wellman's work. However, the Alpine Fault made sense of New Zealand geology and so could hardly be ignored by New Zealand geologists. The crucial question that concerned them was when the movement along the Alpine Fault had taken place. It clearly had to be more recent than the age of the rocks cut by the fault. This narrowed the history of fault displacement down to sometime in the last 250 million years. This was still a very long period, and most geologists tended to want the movement to be as ancient as possible, viewing the uplift of the Southern Alps and associated movement on the Alpine Fault as the product of events that occurred well over one hundred million years ago. By 1955, Harold had found enough evidence to prove not only that these movements had occurred much more recently than this, but that they were actually taking place today. In doing so, he had independently pioneered a whole new field in geology, a field that is central to much of my own research—the study of active deformation of the Earth's crust.

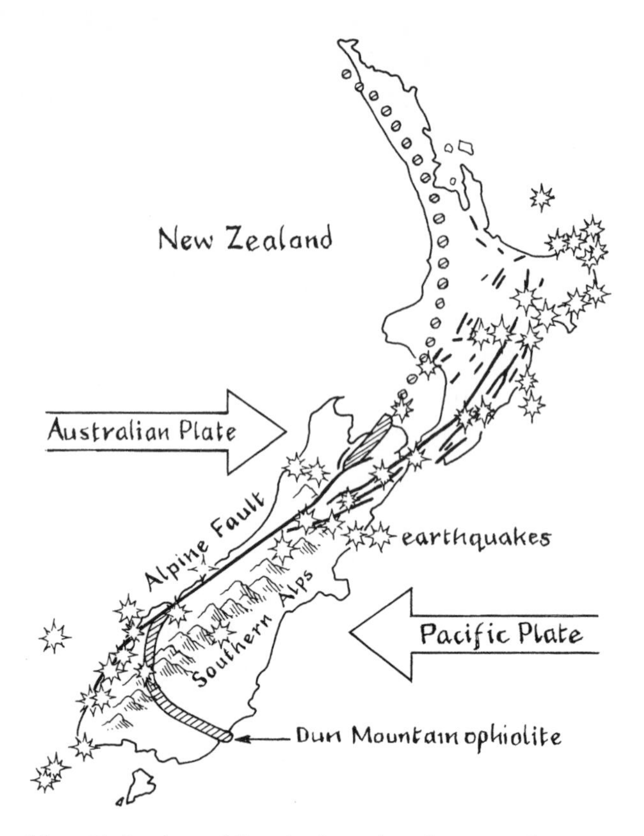

New Zealand straddles the boundary between the great Australian and Pacific plates. The movement of these plates has driven the motions along the Alpine and other faults, triggering the many earthquakes in the region, as well as twisting the rocks to either side and pushing up the Southern Alps.

The key to Harold's work was the recognition that movement on faults occurs during earthquakes—each earthquake is a single increment of movement. New Zealanders had witnessed several of these earthquakes in the short time that Europeans had colonized the islands. In 1855 a magnitude 8-plus earthquake in the southern part of the North Island, just east of the young city of Wellington, had resulted in substantial land-level changes. A wide region, several tens of kilometers across, had been uplifted, locally by as much as four meters. The uplift had immediate practical significance because it had exposed benches of narrow rocky shoreline, which before the earthquake had been passable only at low tide. But now there was easy coastal access for cattle drovers to the recently opened up farming lands in the Wairarapa district, east of Wellington. And it was here, among these farming lands, that a major fault line traversed the landscape, following the edge of a range of hills. Settlers reported that this line had opened during the earthquake as a chasm or rent in the ground.

About a hundred years later, Harold revisited the scene of the 1855 earthquake and searched for signs of movement. He found many features—old river banks or channels, sides of hills, farm fields—that were clearly displaced. It was as though the landscape had been cut by a knife, and the two sides of the cut shifted. In one or two places, where a stream flowed straight across the fault, there was a series of parallel dry channels on the down-dropped side that ended abruptly at the fault. The only sensible explanation was that the dry channels were earlier courses of the stream that had been progressively shifted both sideways and downward in earthquakes, abandoned when the stream switched to a new channel in order to maintain its straight line. There were many other fault lines in New Zealand that had moved like this in historical times. For example, an earthquake in 1888 in the Hope Valley, South Island, displaced a farm fence line about five meters sideways. In 1929 a water channel was cut in two during an earthquake, with one side raised up vertically four meters so that the flow of water was reversed. Harold realized that displacements of several meters, every few hundred years or so, could result in both the colossal 480-km horizontal displacement that he had observed on the Alpine Fault, and the uplift of New Zealand's Southern Alps, in much less than a hundred million years. In fact, he eventually showed that this fault slips, on average, about thirty meters every thousand years, so that the full displacement has occurred in much less than twenty million years.

In 1955 Harold Wellman published his geological vision of New Zealand: a zone of moving faults, almost as wide as the islands themselves, ran right down the length of New Zealand. To either side, great blocks of crust were sliding past each other. Harold had shown how repeated earthquakes over surprisingly short periods of geological time can reshape the continents, resulting in huge movements of the Earth's crust and the raising of a mountain range. But all this brought with it a whole host of new questions. Though, by the late 1960s, it had become clear that Wellman's great blocks of crust were, in fact, two tectonic plates—the Australian and Pacific plates—it remained a mystery why there should be so many fault lines between them. What determines this complicated pattern of fractures? Put yet another way, how and why do rocks break? Rocks certainly break when you hit them hard enough with a hammer. How hard, however, is hard enough, and what actually happens when the rock breaks? And, most importantly for understanding what goes on deep in the Earth, what are the effects of very high temperatures and pressures? I believe that these are questions that go to the very root of understanding how mountains are created.

Rock Crushing

It is not just geologists who worry about crushed or broken rocks. Engineers think about this too. In fact, they are often far more worried than geologists, and it is a problem they have faced for a very long time.

In 1174 the citizens of Pisa in Tuscany decided to build a new campanile for their cathedral. It was to be constructed of white marble, quarried from the Appennine hills and laboriously transported to Pisa. Not long after construction had begun, when the stone masons were laying the courses for only the first floor of the tower, the citizens noticed a disturbing phenomenon. Their tower had begun to list over very slightly to one side. Over the next 150 years or so, as building advanced sporadically, successive stone masons tried to correct for this lean by building straight up from the tilted foundations. Despite this, the tower continued to lean. Today, the completed fifty-five-meter-high tower has a distinct curve in it, and the top is nearly five meters off the perpendicular. The stonework is full of cracks. In desperation, engineers have started excavating underneath the foundations on the opposite side of the tower to the direction of lean. This technique not only has managed to stop the tilting, but has even reversed it so that the top of the tower has now righted itself by about forty centimeters since the emergency work began in the early 1990s.

The behavior of the Leaning Tower of Pisa would not surprise a geologist studying mountain ranges. In simple terms, the ground beneath the tower just could not support the massive concentrated weight of the overlying stone structure. Mountains must also be supported by their rock foundations. And like the Leaning Tower, if they overreach themselves, they have a tendency to fall down. In the 1940s and 1950s, geologists began experimenting in earnest with the pressures inside the Earth, trying to find the conditions under which rocks break—a daunting undertaking because the pressures in the crust at a depth of only a few tens of kilometers rapidly rise to a crushing several thousand atmospheres (I will leave it to your imagination to appreciate the likely pressures near the deep roots of high mountains, at depths of almost a hundred kilometers).

My abiding impression of these experiments is one of leaking hydraulic oil. This oil coats the experimental apparatus and drips onto the floor, forming sticky puddles that the technicians vainly try to soak up with sheets of old newspapers. It is not hard to see why: their experiments consist of squeezing rocks with a heavy-duty hydraulic rig. The rig is constructed so that the two ends of a carefully cut cylinder of rock are placed in a vice operated by a hydraulic ram. At the same time, the sides of the cylinder are encased in a pressurized jacket. Oil, pumped into this jacket, presses in on the sides of the rock sample. The experiments consist of increasing both the pressures in the cylinder jacket and the hydraulic ram—oil oozing out of the straining apparatus—until the cylinder either breaks or changes its shape.

You might be surprised to learn that however great the applied pressures, if the cylinder is squeezed equally in all directions by

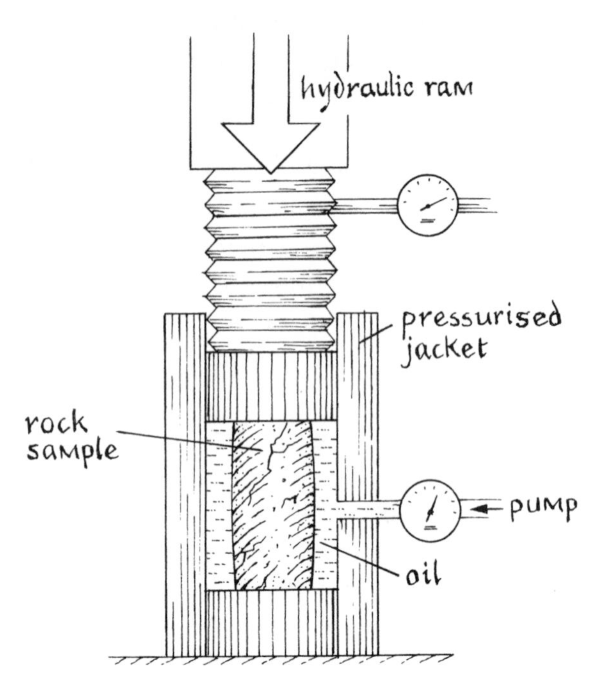

Geologists crush rocks in a hydraulic rig to find out how strong they are.

keeping all the pressures the same, then the rock will never break. What might happen, depending on both the composition and temperature of the rock, is that given a high enough pressure, a new mineral might be created. In principle, diamonds can be made this way by subjecting hot carbon to very great pressures. But the rock will remain intact. However, a small individual adjustment to either the cylinder jacket or hydraulic ram pressures will lead to spectacular results. For example, a typical experiment might be to fix the pressure in the jacket, but allow the pressure in the hydraulic ram to steadily increase above this. With this extra pressure at the ends, the rock cylinder will, at first, compress like a spring, though this springiness is minute compared to, say, a piece of rubber—the moment the extra pressure is taken off, the rock cylinder will bounce back to its original size. However, if the extra pressure continues to rise, the cylinder will eventually start to groan and creak, before suddenly fracturing with a deafening bang. In the right circumstances, the rock will have broken along a perfect plane: an earthquake fault line created in the laboratory. Often, though, the moment of failure is so violent that if the rig was not enclosed, shards of rock would fly out.

This unexpected violence reveals an important aspect of breaking rocks: a rock will break only if it is subjected to uneven pressures over its surface—a result of trying to squash or stretch the rock in particular directions. This is essentially what has happened with the Leaning Tower of Pisa. The underlying foundation is subjected to the enormous weight of the massive stone structure bearing down it, rather like the pressure applied to the end of our rock cylinder in the testing rig. And like the pressure in the cylinder jacket, the push from the surrounding ground has been just too small to prevent the foundation from failing. The tilting of the tower is the direct consequence of this failure, and the extra pressure needed to cause failure is really a measure of the strength of the underlying rock.

Crushing rocks in a hydraulic testing rig provides enormous insight into what happens during an earthquake. Each earthquake is the result of sudden rock failure—it records when a piece of the Earth has broken. The violence, just as in the rock-crushing experiments, comes from the stretchy or springy nature of rocks. But there is one fly in the ointment. The strength of rocks in the Earth seems to be much less than one might expect from experiments in the testing rigs. In 1959, in a paper that immediately became a classic of the geological literature, two Americans, M. King Hubbert and William Rubey, were the first to point out the problem. These geologists had tried to apply the results of the experiments on rock failure to observed faults in the crust. Hubbert and Rubey quickly realized that if the results of the experiments were to be believed, it would be impossible for some of the large inclined faults that had been discovered in the crust to exist. In simple terms, the experiments suggested that it would require unrealistically large forces to move blocks of crust along such large faults—the rocks would bust up into a myriad small fragments long before there was any slip on the big fault. To resolve this paradox, Hubbert and Rubey needed to find a lubricant that would allow the big faults to slip much more easily.

It soon became obvious to Hubbert that this lubricant is water. He had worked in the oil industry and knew well that whenever you bore deep into the ground, you will find water. It permeates cracks and fractures in the crust. However, it is also under very high pressure. Hubbert suggested that it is this pressure that effectively lubricates the fault by pushing the two sides apart, allowing the rocks to slip past each other much more easily. In this way, the outer part of the Earth is fundamentally weakened. Returning to that medieval monument to instability—the Leaning Tower of Pisa—it has become clear that its problems also stem from the presence of water; the ground beneath the tower is saturated in it.

How to Squash a Fudge Cake

You might be asking yourself how all this concern with squashing rocks relates to the Andes. Our own work, completely independent of these rock-crushing experiments, quite easily established that these mountains are just a heap of broken rock. Some geologists have even made the analogy with sand—the Andes are a pile of sand. So what? Well, the earthquakes that periodically shake the region, like the one that destroyed Aiquile in the Bolivian Eastern Cordillera, tell us that this heap is still shifting around and cracking. So, straight off, we know from the earthquakes that at least some parts of the mountain range are growing today, getting just a little bit higher as each year, decade, or century passes. But consider this. Leaving out the trail of deep earthquakes in subduction zones, seismologists have found that almost all other earthquakes occur at depths shallower than 40 kilometers, and usually no more than about 15 km. But we know that the roots of the mountains extend down much deeper, to depths up to 70 kilometers, and below that there are mantle rocks. So what is happening here? Why are these rocks not cracking and fracturing as well? And this is where the testing rig experiments become really interesting.

Everything I have talked about so far applies to relatively cold rocks, at temperatures up to a few hundred degrees centigrade. Under these conditions the rocks are said to be brittle. But warm the rocks up and there is a drastic change: they will start to be soft and pliable. This is a bit like the difference in behavior between a hard bar of chocolate, straight from the refrigerator, and a soft one that has been lying in the sun for a while. Try following this recipe: heat a cylinder of rock up to more than about one-third of its melting temperature in a cleverly designed furnace, then apply uneven pressures and watch it slowly start to creep or flow, bulging out sideways. This sort of behavior is called plastic or ductile rather than brittle. In fact, as the rock gets hotter—though it is still well below its melting point—it flows more and more easily as a very sticky fluid. This fluidlike behavior of a hot, solid rock may come as a surprise, and, at first, seem paradoxical; how can something that is solid flow? This is a question that deeply puzzled me when I first studied geology in college. Yet, if one thinks about it, there are many, much more familiar substances that do just this. You may not consider warm chocolate in the same category as rocks. But what about metals such as lead, copper, or gold? These can be easily bent or drawn into a wire—in effect, flowing—while still remaining solid and far from melting.

The changeover in the Earth between the ductile and brittle behavior of rocks is called the brittle-ductile transition and corresponds to a rock temperature of about 350°C—the temperature at which ductile behavior starts to occur in the hydraulic rig experiments. We can now begin to make sense of the earthquakes. Measurements in boreholes, drilled by oil companies in the Andes, show that the temperature generally increases about 25°C for every kilometer of depth. I once had a firsthand experience of this increase in temperature when I entered one of the Bolivian silver mines in Potosi. We walked for over a kilometer along tunnels that had been bored far into the Cerro Rico, effectively traveling deep underground. The air got hotter and hotter, and I was soon dripping with sweat, despite the fact that the outside air temperature at the mine entrance was below freezing. So it is clear that temperatures as high as 350°C could be expected at a depth of much less than 20 km, more than enough to "soften" the rocks into a warm, toffeelike behavior. No wonder there are no earthquakes at these depths below the Andes.

Let me use the behavior of a gooey fudge cake with a white sugary icing—the sort of cake you should avoid if you are on a diet—as a way of summarizing how the Andean mountains with their deep roots must respond to being squeezed. Squeezing of the whole cake will certainly result in cracking of the "brittle" icing, and flow of the underlying "ductile" fudge. In the same way, as we have seen, squeezing in the Earth will cause the top few tens of kilometers to fracture—the behavior of the icing— where the rocks are cold enough to be brittle. And, at greater depths, the rocks are now hot enough to be plastic or ductile and flow—though they are still very sticky—rather like the gooey fudge part of the cake. In fact, even before considering the rock-crushing experiments, we had tacitly acknowledged this flow when thinking about the roots of mountains floating on the underlying fluidlike mantle—Airy's iceberg model for the crust (see chapter 6).

I don't wish to confuse you by making things even more complicated, but the simple fudge cake is not quite enough to be a good analogy. To see why, we need to consider what part of the Earth is actually involved in mountain building. It is clear that the crust is important—it is the deep roots in the crust that underpin the mountains. But what about the underlying mantle rocks? Mountain building does not just stop at the base of the crust—much more of the Earth's interior is being squeezed when the mountains are pushed up. Geophysicists agree that at depths in the Earth greater than 100 kilometers or so, the mantle rocks are far too hot and runny to play much of a role—they just flow away the moment they are squeezed. This leaves us with the outer shell of the Earth, a region that is relatively cool and strong, called by geologists the lithosphere (litho means rock in Greek).

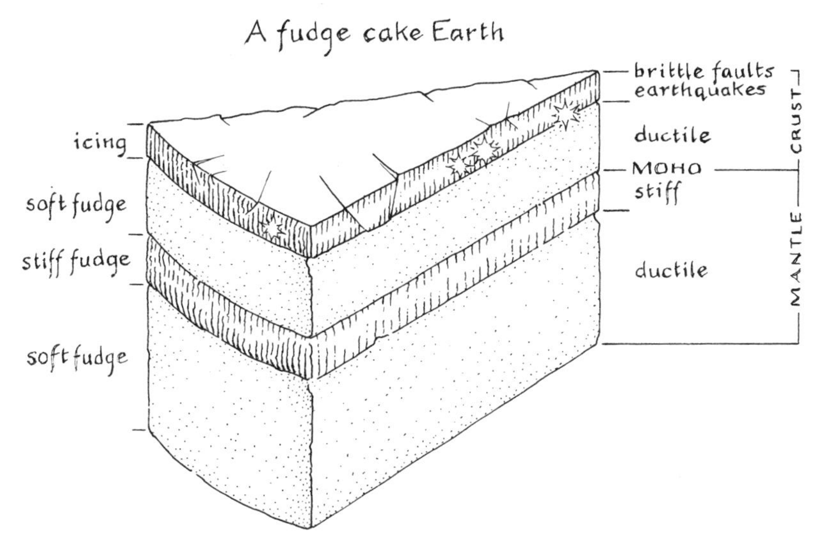

The outer strong portion of the Earth, called the lithosphere, is a bit like a fudge cake. The top of the crust is the brittle icing, cracking during earthquakes. Below is the fudge part, where the crust is hotter and the rocks flow. Even deeper, in the uppermost mantle, the rocks have a different composition and may be slightly stiffer fudge. As the mantle becomes hotter with increasing depth, it flows much more easily, again like soft fudge.

Returning to the fudge cake analogy, it is now clear that it is really describing the lithosphere, and that this is not a single substance but contains bits of the crust and underlying mantle. These are layers of the Earth that have different compositions. This really boils down to a difference in minerals, and so we need to be much more precise about the compositions of the crust and mantle. The crust beneath the continents is the stuff of geological maps, made up predominantly of sedimentary rocks and granite that both contain abundant quartz and feldspar—the typical minerals, for example, of the sand grains on the banks of a river. Unfortunately, because geologists do not normally get to see the underlying mantle, its composition is a bit more difficult to work out, though there are a number of important clues. To begin with, blocks or nodules of a green rock, composed mainly of olivine, are occasionally brought up from great depth in volcanic eruptions. Second, experiments have shown that these nodules will melt to form basalt. And, as basalt is the commonest type of volcanic lava, there must be a lot of rock similar to the green nodules beneath the crust. We can therefore safely say that the mantle is made up mainly of olivine.

The minerals quartz and olivine determine, to a large extent, the squashing behavior of the lithosphere. Experiments with these minerals in a rock testing rig have shown that, at high temperatures, olivine offers more resistance to squeezing than quartz. These differences make the lithosphere, overall, a much more unusual material—a cake with several layers in it. Understanding in detail how this sort of lithosphere responds to forces is a difficult problem—something geologists have still not fully solved. One might ask whether this problem is important anyway. The answer is that it is very important if one wants to know, for example, why, when, and where earthquakes occur; or what controls the overall shape and height of mountain ranges; or where mountains, or large depressions, are likely to form on Earth or other planets.

Imagine being blindfolded in a cake-testing laboratory, and just listening to the results of a series of experiments involving squashing of different types of cake. If you heard a "crunch" noise you might deduce that the cake had an icing that was thick and hard—the crunch was the sound of it cracking, and this is where the cake's strength lay. But perhaps you heard just a soft squish? This might suggest to you that the icing was thin and weak, and the experimental response of the cake, as a whole, to squeezing or squashing was really dictated by whatever lay beneath the icing. But moderately thick icing and various layers of fudge with different consistencies would give the cake properties that are somewhere in between and more difficult to visualize (especially if you are blindfolded!). However, when geologists first started thinking seriously about this problem, it seemed that the lithosphere really did have a thick and hard brittle "icing" and was pretty soft and runny underneath.

Continents That Crack

In the early 1970s, Dan McKenzie, at Cambridge University, began to study the earthquake-prone regions of the Middle East and Mediterranean. McKenzie displayed a dazzling brilliance in his ability to see the essence of a geological problem and present it in a way that was amenable to mathematical analysis. He had already played an important role in the development of the theory of plate tectonics, when, in 1967, he had, together with a colleague, Bob Parker, published a paper setting out many of the geometrical consequences for an Earth covered with moving rigid plates. McKenzie was, therefore, well aware of the significance of faults in the crust, such as those recognized in northern Syria and Turkey that linked up with the Hellenic trench south of the Aegean. At first, McKenzie interpreted this region in terms of the boundaries of numerous small tectonic plates—micro plates—caught up between the great European and African plates, rather like slabs of broken sea ice, jostling together between two large, floating ice sheets. This view of the continents, cracking up into micro plates at the boundaries of the major tectonic plates, was given a new lease on life by a Frenchman, Paul Tapponnier, when he started to examine satellite images of Central Asia, including the great ranges of the Himalayas and the Tibetan Plateau.

The satellite images were crossed by knife-sharp lines along which the grain of the bedrock abruptly changed. Some of these lines extended for hundreds or even thousands of kilometers. Tapponnier identified these lines as gigantic faults in the crust, on a scale similar to Harold Wellman's Alpine Fault in New Zealand or the San Andreas Fault in California. The most spectacular example in Tibet is the Altyn Tagh Fault, which extends for over a thousand kilometers in an almost dead straight line in northern Tibet, along the southern margin of the Tarim Basin. Again, like the Alpine or San Andreas faults, the detailed expression of these faults, as they crossed the varied landscape of Asia, indicated that they were principally strike-slip faults along which the crust had slid horizontally. The overall pattern of faulting suggested that Central Asia was divided into several large micro plates or blocks.

Tapponnier's observations created a great splash in geological circles, and, in the mid-1970s, they were written up—coauthored with the American geophysicist Peter Molnar—in the prestigious journals *Science* and *Nature*. The widely read articles concluded with a truly grand picture of movements in the continents: fragments or blocks of Central Asia had slid hundreds—sometimes, thousands—of kilometers along huge strike-slip faults into what today is Southeast Asia. All this, they believed, was because the region was caught up in the collision between two tectonic plates—India in the south and Eurasia in the north—cracking into huge blocks. Plate tectonic theory predicted that this collision had started about fifty-five million years ago, and since then, India has pushed some two thousand kilometers northward into Asia. This relentless northward movement, according to Tapponnier and Molnar, had forced aside the region in front, allowing it to escape sideways and out of the way, squeezed out like so many fruit pips, along giant strike-slip faults. It was as though India was a gigantic icebreaker, making its way through a frozen ocean and cracking and pushing to one side large rafts of ice. More fundamentally, the evidence for fracturing on this scale seemed to show—returning to my fudge cake analogy for the lithosphere—that the top layer of icing offers considerable resistance to squeezing and plays an important role in controlling the cracking of the whole cake into several large fragments that slide past each other.

Tapponnier and Molnar's idea was taken up enthusiastically by many geologists. Some were puzzled, but their quarrel was not with the theory of plate tectonics. It was just that they found it hard to see how sideways escape of giant blocks of crust could explain the existence of the vast ranges of Tibet and the Himalayas—the highest regions on Earth. And they doubted that the brittle crust was really quite so strong. Something was clearly missing.

Continents That Flow

Dan McKenzie, who had been one of the first to try and make sense of fault movements in the continents, was beginning to have second thoughts about some of his old ideas. Working together with a young geophysicist, Philip England, he decided to tackle the problem by taking a completely different approach. I have to confess that though I was, at the time (the early 1980s) a student in the same university department, I was completely unaware of McKenzie and England's work. In marked contrast to Tapponnier and Molnar's theory, they had decided to investigate the consequences of treating the continental lithosphere—in particular, the Eurasian lithosphere—as a fluidlike material instead. Put another way, they wanted to analyze the behavior of the gooey cake on its own, without any brittle icing.

This radical approach was prompted by the fact that, by now, seismologists were detecting earthquakes—a sure sign of rocks breaking—almost everywhere in large tracts of the continents, especially where there were mountains. Clearly, the rocks were breaking in many more places than just along the giant faults that are so prominent on the satellite images. Going on over time, this would have fragmented the continental crust in Central Asia, and much of the mountainous region extending farther west to the European Alps, into much smaller pieces than had been previously thought. Indeed, this might even have been anticipated, since many geologists believed that the water deep in the crust was under high pressure, acting as a lubricant that could fundamentally weaken the crust.

The basic thinking behind treating the lithosphere as a fluid is the idea that it has no long-term strength. As we have already seen, strength to an engineer or geologist means the ability to withstand unequal forces or pressures without breaking or distorting. From this point of view, a fluid has no strength at all, though, of course, its stickiness can offer considerable resistance to being squeezed. However, given the slightest unequal pressure, the fluid will flow.

The speed at which it flows will depend on both the actual pressures and the stickiness or stiffness of the fluid. For example, a mound of sticky syrup—molasses or treacle—in a container is subjected to the vertical force of gravity but is unsupported around its sides—clearly, the resultant pressures in different directions are unequal. As a consequence, the syrup will flow outward. Only when the syrup eventually fills the container with a level surface will it stop flowing. At this stage, every bit of syrup will be subjected to a uniform pressure in all directions, exerted by the weight of the syrup pressing in on itself. The lack of strength means that gravity plays an important role in the flow of fluids—it is the force of gravity that causes the syrup to flow out and fill its container, and it will do the same for fluidlike continents.

McKenzie and England applied their new idea to the northward advance of the Indian continent into Eurasia. Much like Tapponnier and Molnar, they considered India as a gigantic piston, relentlessly pushing forward, but now there was an important difference. Instead of the region in front, in Eurasia, behaving as hard "icing" or even ice, breaking up into a few large micro plates that were being squeezed out sideways—Tapponnier and Molnar's idea—it was being treated as a fluid; India was no longer ploughing through thick pack-ice, but very sticky syrup instead. We could imagine Eurasia as a tank of fluid. The push of the Indian piston into Eurasia is like forcing more fluid through an inlet in the side of the tank; the rate at which this fluid is pumped in, and the width of the inlet, simulate the speed and size of northward-moving India.

Initially, the fluid in the imaginary tank has a level surface. However, as more fluid is pumped in, a mound starts to build up in front of the inlet—in effect, piling up in front of advancing India. This is merely because the fluid is entering the tank faster than it can flow away under the force of gravity—India is advancing faster than the fluid in front can flow out of the way. The rate at which the mound does flow away depends on its stickiness and height. So eventually, some balance is achieved and a plateau of fluid develops around the inlet that maintains a more-or-less constant height, but just gets wider and wider as more and more fluid flows through the inlet—at

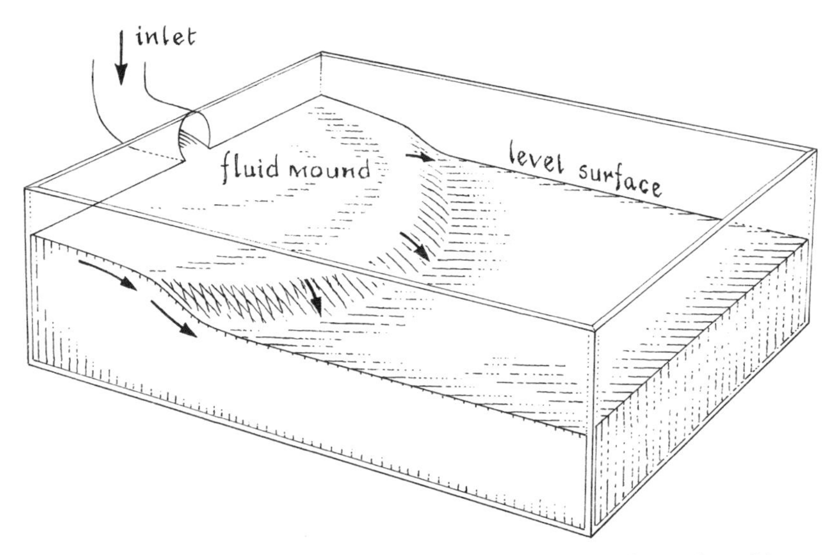

The great mountain ranges of Asia have features that can be explained by a fluidlike Earth. Pouring sticky syrup into a tank creates a mound if the fluid enters the tank faster than it can flow away. The shape of the mound, together with the radial pattern of flow at its edges, is similar to the shape and "flow" of the high Tibetan Plateau.

its edges the flow tends to be radially outward. The plateau can be elevated by increasing the rate of inflow or stickiness of the fluid, but strengthening the force of gravity will lower it. For example, fluidlike mountains on Earth will be lower than those on the Moon or Mars, where the force of gravity is less.

This plateau or flat-topped mound of fluid seemed to McKenzie and England to be strongly reminiscent of the great Tibetan Plateau, which forms a vast region in Central Asia, several thousand kilometers across, at a general elevation of roughly 5 kilometers above sea level. Also, the predicted radial pattern of flow seems to be found in the observed squeezing around the edge of the Tibetan Plateau. It was beginning to look as though McKenzie and England's simple analysis could explain some of the features of the great ranges of Asia. But the fundamental problem with this model, obvious to many geologists working in the region, remained. The brittle layer of "icing," with its major fault lines, had been completely ignored.

Having Your Cake and Eating It

Again, it was Dan McKenzie who provided a possible solution to how the Earth's surface could both break up along major faults—the situation that seemed to exist in Central Asia and also farther west in the Middle East and Mediterranean—and behave, overall, as a fluid. In 1983 he published his ideas in a paper written together with James Jackson, who had worked with McKenzie on the pattern of faults and earthquakes in the Middle East.

I attended the departmental seminar in Cambridge when Dan first presented his ideas about faults and fluid flow. He illustrated his talk with some wooden models he had made—the craftsmanship involved in making these intrigued me. Interestingly, in this respect he is very similar to that New Zealand pioneer of recognizing faults in the landscape, Harold Wellman. Harold always resorted to simple geometrical constructions, or models made out of cardboard or wood, to illustrate his ideas about how faults worked. However, much of Dan's talk was clouded by dense mathematics, which I did not understand at the time. When I finally came to read carefully and understand his work a few years later, it turned out that I was one of the few geologists who had bothered to do so—or so Dan told me when he reviewed a paper I wrote on a particular aspect of this work.

Let me illustrate with the fudge cake analogy how the surface of the Earth can appear both to be a fluid and to break up along fault lines. What happens is that the thin topping of icing cracks and moves with the underlying flow of the rest of the cake. This would be rather like pieces of wood floating on top of flowing water, or fractured floating ice, shifting and jostling as it is carried along by the variable sea currents. The crucial point is that the motion of the icing is driven by the underlying flow of the cake. Dan imagined that the fragments of icing were like tops or buoys, drifting and

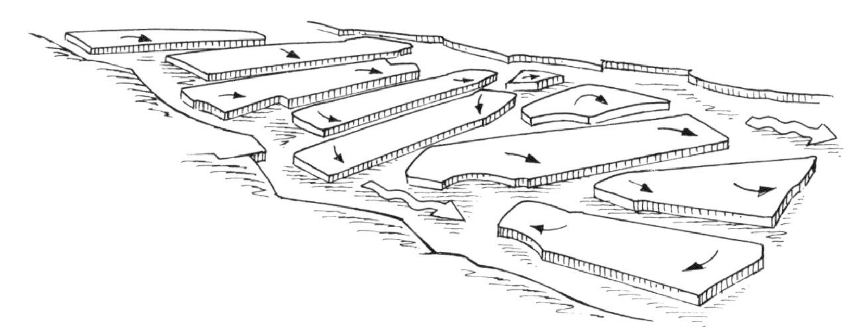

The motion of shifting and jostling fragments of ice, floating in a current of water, may be a good analogy for the behavior of the Earth's crust where it is broken up into blocks along moving fault lines. This makes sense if the motion of the brittle crust is really controlled by a fluidlike flow in the underlying ductile rocks.

spinning at the same time. He could also predict how fast these tops or buoys would do this. Blocks of crust in the Aegean, or Middle East or New Zealand, should have spun around a quarter of a turn or more, like so much flotsam and jetsam in an eddy of a river, in a relatively short period of geological time—not more than a few million years. And this has now been confirmed by measurements using the magnetism in rocks: parts of New Zealand have swung around not far off a quarter of a turn, drastically reorientating the axis and coastline of North Island.

It seems that not all of the Earth's surface is covered by rigid tectonic plates; there are significant areas, especially in the continents, where there is no sign of these plates at all, but a swirling, fluidlike flow. But not all geologists are convinced: they believe that however much one wriggles, scientifically speaking, one cannot escape the fact that the crust breaks up into large fragments along only a few major faults, the faults that can clearly be seen, extending for hundreds of kilometers—the Alpine Fault in New Zealand, the San Andreas Fault in California, or the Altyn Tagh Fault in Tibet, to name but a few.

The champions of the fluid theory argue that wherever there are large fault lines in the continents, there are likely to be many much smaller ones, more difficult to detect. Though the big faults may be long and prominent, they are not necessarily the most important in terms of the amount of movement that has occurred along them. Can you imagine trying to account for every fracture in the icing of our fudge cake after it has broken into crumbs of sugar? Surely, only the big breaks would catch your eye. What was urgently needed was some way to accurately measure these movements.

Measuring the Flow

In 1984 Dick Walcott, who had been working as a New Zealand government geophysicist, visited Cambridge on sabbatical leave before taking up a new appointment as professor of geophysics at Victoria University in Wellington, New Zealand. At this time I was writing up my doctorate thesis on some of the oldest rocks on Earth, exposed in the small southern African kingdom of Swaziland. Dick's Cambridge visit had been partly sponsored by my thesis supervisor, Alan Smith. Alan had arranged for Dick to hold a St John's College visiting research fellowship.

One afternoon, as I leaped past Alan on the main staircase in the department, taking two steps at a time in my urgency to get to my office, he asked me if I had any plans for my future. He told me that Dick Walcott was looking for a postdoctoral researcher to work with him in New Zealand. Would I be interested? A picture of green fields, snow-capped mountains, and sheep immediately appeared in my mind. The next day at morning coffee, Dick Walcott took a seat beside me and asked about my thesis research. I could see his interest in me begin to fade as I described how I had been studying these incredibly old rocks in Africa, trying to work out what the very early Earth was like. Dick remarked that there were some similarly ancient rocks in the Fiordland region of South Island, but they were very difficult to get to and often obscured by thick bush. New Zealand was not the right place, he concluded, to study the early Earth. It seemed that I had lost my chance of going to New Zealand.

A few days later, Dick Walcott gave a talk in the department on his New Zealand research. The first slide, to put us in the mood, was a picture of a big woolly sheep staring out from a bright green field, with a clear blue sky overhead. The talk, though, was on a completely new topic for me. Dick described how the work of land surveyors in New Zealand—straddling the very region where Harold Wellman had found his many fault lines between the Pacific and Australian tectonic plates—could be used to show that the country was continually moving today. These land surveys had first been made in the 1880s when surveyors from the British Survey in India had started to fix the newly colonized country with the greatest precision that was possible at the time. They were designed to stand for all time as a framework for maps of the country. The technique was to climb to the top of a hill and sight on several neighboring hilltops, measuring the angles between the sightings. The lines of sight between points built up a pattern triangles that form the backbone of any map—for this reason, the hilltop markers are called triangulation or trig points. The instruments they used could theoretically measure the angles between lines of sight to an accuracy of a few parts in a million. This is equivalent to fixing the position of a point over a kilometer away to the nearest millimeter.

Over the years, as more detailed maps of New Zealand were needed, subsequent surveys expanded on the early work. To the great embarrassment of the surveyors, each new survey never quite agreed with the earlier ones. It seemed, at first, that the early surveyors were not as accurate as they thought in fixing positions on a map. It was Harold Wellman in the 1940s who first suggested that the real problem was not the early surveyors' accuracy, but New Zealand itself, which was constantly shifting and changing shape because of the movement along fault lines. Dick Walcott together with Hugh Bibby, at the New Zealand government Geophysics Division, had followed up this initial idea by systematically comparing triangulation surveys made in the 1880s with those made up to a hundred years later. They did this by comparing the angles measured in the triangles during different surveys. They could estimate the accuracy of the surveys by adding up the angles in each triangle—these should always come to 180 degrees if there are no errors at all in the measurements.

In this way, they found that the angles in the triangles changed in a systematic way from one survey to another. The only explanation was that New Zealand had been distorted during the last hundred years—a steady movement of the landscape, consistent with the general pattern suggested by the slip on the many fault lines in the same region. In other words, the surveyors had directly measured the flow of New Zealand. The rate of movement can be calculated by looking at the total relative shift across New Zealand, as the western part of the country slides past the regions farther east. The change in the angles of the triangles during the last one hundred years showed that this total shift was, on average, about forty millimeters each year—near enough to the motion of the Pacific and Australian plates given by the theory of plate tectonics.

One potentially confusing aspect of Dick Walcott's work is that the movement he had measured was much greater than all the observed slip in earthquakes along the fault lines in the country during the same period—in fact, there had been very few earthquakes. To see why this might be the case, it is necessary to recall what happens between earthquakes. We have already established that an earthquake is the result of sudden fracturing of rocks along a fault. Before the earthquake the forces slowly build up in the crust, driven by the motion of the tectonic plates. In the process, the rocks are stretched or squeezed like a piece of elastic or rubber over a wide region. Eventually, the elastic reaches its limit of distortion and snaps—this is the earthquake, when the elastic stretching or squeezing is converted into sudden slip on the fault. So the triangulation surveys were mainly recording the widespread elastic distortion of New Zealand between earthquakes. The important result is that the movements vary smoothly throughout New Zealand, as though it were flowing. Walcott thought that he might also be measuring the motion that was taking place much deeper down, in the ductile part of the lithosphere, where the rocks are hot enough to really flow like a fluid; the triangulation surveys might also be revealing a true fluidlike flow.

I remember that there was a murmur of appreciation from the audience after Walcott's talk. Dan McKenzie, from his table-top perch at the back of the room, remarked that Dick must be the first person to reveal this widespread pattern of flow so convincingly—"You have done what I did not think was possible." Next day, when I had a chance to speak to Dick, I enthused about his talk. He was witnessing, I said, the process of earth movements. In Swaziland, I had been studying the effect of earth movements that had occurred an unimaginably long time ago. I now wanted to visit the places where they were happening today, observing how rock strata fracture and fold. Dick seemed impressed by my Damascus-like conversion. In that case, he said, New Zealand could be the place for you. And so, with his help, I applied for a New Zealand government three-year postdoctoral research fellowship, studying the active distortion of the New Zealand crust. During the ensuing months, as I struggled to finish my thesis, that picture of the woolly sheep in a green field was constantly in my mind.

I eventually went to New Zealand in 1985. Here, I learned for myself how to think of the movements in the crust in terms of the flow of a fluid.

The Flow of the Andes

When I first started to study the Bolivian Andes, I intended at some stage to test the idea that these mountains too had flowed up like a fluid. But it was not until I had been working on the Andes for several years that I was in a position to do this. The main problem was simply finding a way to measure the movement. There were no suitable old surveys like the ones Dick Walcott had used in New Zealand. The only solution was to rely on some other technique. Today, walkers, sailors, and even car drivers routinely determine their position anywhere on Earth to within about six meters, using a small, handheld Global Positioning System (GPS) receiver. This is certainly good enough for most people. Except geologists.

At typical rates it would take more than two hundred years for the crust within a mountain belt to move as much as six meters. Clearly, we need a much more accurate system of determining position. We need to be able to calculate positions to the nearest few millimeters. And it turns out that it is possible to do this using GPS. Rather than making a quick measurement with a handheld instrument, we need to use a more sensitive receiver and make measurements at the same point for several hours, taking full advantage of all the information transmitted from the satellites in space. In this way, measurements can be made to the necessary accuracy. If the measurement is now repeated a few years later at exactly the same spot—this has to be carefully located with a survey peg—then the change in position of the spot can be calculated. Dividing this change in position by the time interval between surveys yields the speed of the point, typically several millimeters to a few tens of millimeters per year.

Between 1994 and 1996 an international team of surveyors lead by Edmundo Norabuena, a Peruvian geophysicist, used GPS to measure changes in position of points in the Bolivian Andes relative to the rest of South America. Unfortunately, they measured this motion for only a few points; not enough to get a good picture of the pattern of flow. These results were published in the journal *Science* in 1998. When I saw their results, I wondered if it would be possible to somehow combine them with other clues to the movements in the crust that I had found myself. In this way I would be able to produce a much more complete picture. For example, our study of the remnants of a peneplain in the Eastern Cordillera (see chapter 6) had shown that there had been no squeezing of the rocks in this region during approximately the last ten million years. Yet, smoothing out the crumples in the Sub-Andes, farther east, had revealed substantial squeezing. Measurements of the magnetism in the rocks (see chapter 7) showed that blocks of crust throughout the Bolivian Andes had spun around up to fifteen degrees during the same period—counterclockwise in the north and clockwise farther south. In addition, the pattern of earthquakes and faulting revealed something of the direction in which blocks were moving, but not how fast.

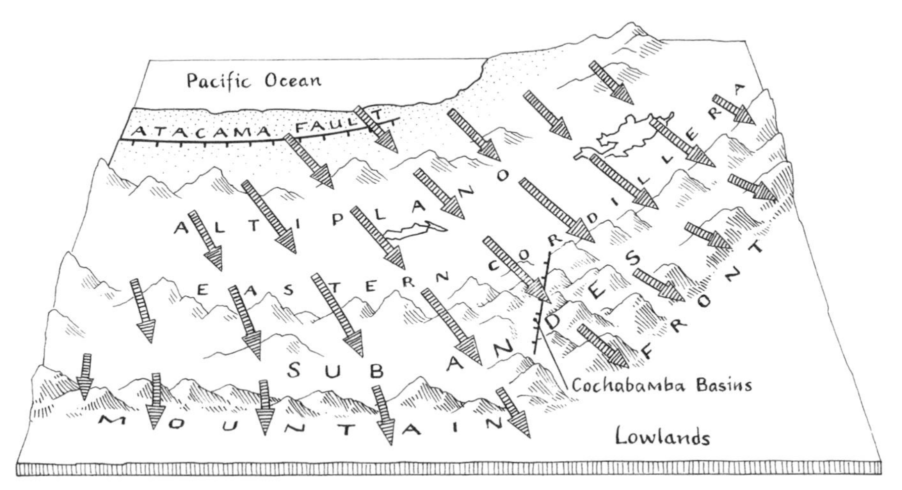

Precise measurements of the motion of rocks in the Bolivian Andes suggest an overall pattern of flow toward the eastern lowlands. The Atacama Fault in the west and the Cochabamba basins farther east—places where the crust is splitting apart—may all be expressions of the tendency of the mountains to flow downhill.

I remember that the solution to this problem came to me while I was giving a lecture. It really was quite simple. I could turn the whole problem on its head. If I knew the pattern of flow for the Andes, I could calculate the squeezing, rotation, and direction of fault movement anywhere I wanted. So all I had to do was try different patterns of flow and see which one agreed with my observations. I had as a starting point the GPS measurements made by Norabuena's team. It was just a matter of trial and error, filling in the gaps. This sort of approach would have been impossible a few years ago. Modern computers have changed all that. The final computer program took several months to develop and involved several thousand lines of computer code. The first time I ran it, using all the available data, I was excited to see what the computer would come up with. And to my amazement, the pattern of flow was really quite simple and smooth—a good sign that my method actually worked.

As all motion is relative, the movement of the mountains must be measured relative to something. I had decided to calculate the motion relative to the lowlands of Bolivia, east of the Andes in the Amazon jungle. From this viewpoint, the mountains everywhere appear to be flowing toward the lowlands. The most distant ranges on the extreme western edge of the Andes are moving the fastest—roughly ten to fifteen millimeters or half an inch each year (at this rate, it would take well over ten thousand years to walk to your local shop). Ranges closer to the lowlands are moving slightly slower. But the most dramatic drop in speed occurs across the Sub-Andes on the eastern margin; by definition, the lowlands themselves are not moving. One of the most sensible features of the flow is its direction. On the western margin of the Andes, the direction of flow is essentially the same as the direction in which the two great tectonic plates—the Nazca and South American plates—are moving together. This direction has recently been independently measured with great accuracy, also using GPS. The Nazca plate moves toward South America in a direction that is slightly north of east, on a bearing of about 080°.

The flow in the Bolivian Andes is an extraordinary clue to the origin of these mountains, at least over the last few million years. It is clear that as the Nazca plate inexorably glides toward the South American plate, it bends down in the trench offshore and slides underneath the Andes into the Earth's interior. However, this is not the whole story. The GPS measurements reveal a total relative movement for the plates of about seventy millimeters (or three inches) per year. The flow of the Bolivian Andes shows that only a small fraction of this movement—less than a fifth—is absorbed in the mountains themselves. It is this fraction that results in the Andes being squeezed and pushed up. The remaining four-fifths of the plate movement— about fifty-five to sixty millimeters per year—is taken up in the subduction zone, offshore, where the Nazca plate simply slips—without much fuss—beneath the western margin of South America. This situation is a bit like a badly slipping car clutch, which transmits only a small proportion of the engine's power to the wheels, or in our case, to pushing up the Andes. The conclusion must be that subduction zones are surprisingly inefficient engines for mountain building.

There is another remarkable aspect to the pattern of flow. This becomes clear if one thinks about the prominent bend of the Andes in Bolivia. The mountain range swings around from a general north-west direction in northern Bolivia to nearly north-south, in southern Bolivia and northern Argentina. The flow suggests that the mountains are moving slightly faster in the angle or knee of the bend, compared with the thigh and calf of the Andean leg farther north and south. Over time, the knee has advanced farther toward the lowlands than other parts of the leg, a situation that could be explained if the leg is pivoting at the knee. We seem to be observing in the pattern of flow a huge mountain range changing shape as it bends like a hinge.

Mountains That Flow Downhill

At this point we need to reexamine carefully what we actually mean by the flow of the mountains. Despite all our talk of this flow, and, indeed, the attempts to measure it, we have not conclusively proved that it is the motion of a real fluid. We can, in fact, describe many materials as flowing. For example, we might say that sugar emptied from a bag is flowing. But sugar is clearly not a fluid; it is made up of many small grains. In this case, the "flow" describes the overall motion of the many small grains. In the same way, the flow of the mountains, at least at the Earth's surface, is the overall motion of many blocks of solid crust. Going right back to the original question I asked about the behavior of a fudge cake when squeezed, we need to demonstrate that we can effectively ignore the brittle icing and regard the motions of the many blocks as being driven solely by the flow in the underlying fluid fudge.

There is a way of doing this. If the cake, as a whole, really is flowing like a real fluid, then there will be a tendency for both the brittle icing and underlying cake to flow downhill, strongly controlled by the force of gravity. And so the question becomes whether we can see this tendency in the surface flow of the mountains. If we cannot, then we would be forced to conclude that Dan McKenzie and Philip England's fluid theory is wrong. It turns out, however, that there are, indeed, signs that the mountains are flowing away. On the eastern side of the Bolivian Andes, the movement tends to point in the direction of steepest slope, straight toward the lowlands.

This pattern of flow also helps to explain two features of the Andes that I had encountered on my very first visit to the region. When I first visited Cochabamba, in the Eastern Cordillera, I was struck by the evidence for a great rift in the mountains. It is now clear that this rifting is really an inevitable consequence of the flow of a fluid mountain range spreading out in the core of the great bend or knee of the Bolivian Andes, forever seeking the direction of steepest regional slope on its eastern edge. This way, the crust is pulled apart, opening up the rift of the valley of Cochabamba. Much farther west, the Atacama Fault, which runs along the length of northern Chile, is also a sign of this, this time along the Pacific edge of the Andes. Here, the crust of northern Chile is being pulled apart as the coast literally falls away into the deep ocean trench offshore. Borrowing more of Philip England's ideas about the flow of fluid mountains, I found I could go one step farther down the road of fluidity by using the rate of flow to actually work out the average viscosity of the Andes, or how runny they are—a colossal 100 billion trillion poises (poise is a unit of viscosity). Molten lava has a viscosity of only about a hundred poises, and water is about one-hundreth of a poise, so rocky mountain ranges are clearly pretty stiff stuff, but fluidlike nonetheless.

A Final Hurdle

There is a catch to all this. Though it is true that the pattern of flow we can observe at the Earth's surface has many of the properties of the flow of a real fluid, such as a tendency to flow downhill, there is good geological evidence that something is peculiar about this flow at depth. For example, I described in an earlier chapter the evidence that the whole eastern side of the Andes was sliding up a gigantic, gently inclined fault, overriding the rigid Brazilian Shield. At first sight, this sort of movement does not look like the flow of a fluid.

This apparent contradiction really stems from the fact that we have oversimplified the problem. As we have already acknowledged in our cake model, the part of the Earth that we are talking about—the lithosphere—is not uniform but has layers with different strengths. In addition, the evidence for the high strength of the Brazilian Shield, in the heart of South America, points to marked lateral variations in the stiffness of the lithosphere. So we are really talking about the flow of not just one fluid with a uniform viscosity, but a rather complicated combination of fluids with varying viscosities. It is actually possible to simulate inside a computer how such a complicated combination of fluids would flow for any specified set of conditions, and a number of geophysicists have now made these calculations. Their work is beginning to show that if we squeeze this type of fluid in the way suggested by the pattern of flow of the mountains, it will start to behave in detail much more like our geological observations. The reason for this is that the flow tends to be focused into the runniest parts of the fluid. So if we think of the Brazilian Shield as a fluid that is, say, ten or may be a hundred times stiffer than the Andes themselves, then it will actually remain more or less intact—forming a rigid tectonic plate—like a slab of cool toffee in a pan of warm sugar syrup. And the mountains, like the warm syrup, will flow over it.

Finally, if we step back from the mountains, we find their essential fluidlike nature is revealed in their overall shape. In Peru, Bolivia, and northern Argentina, the Andes contain a broad central plateau, about 4,000 meters high and remarkably level. It is hard to imagine how anything but a fluid could settle out, almost like a spirit level, to form this level plateau—like the great plateau of Tibet—among all the upheavals of the Earth's surface that have created the mountains in the first place. So once again we are forced to contemplate the fluidity of our planet, coming about as far as we can with this idea. We have watched the Andes flow, albeit like a rather complicated fluid. We have even managed to get some idea of their runniness or viscosity. And we now understand precisely the role they play in accommodating the convergence of the two tectonic plates. But perhaps the most important result of treating the Earth beneath mountains as a fluid is that we can use the resultant flow to work out the forces that drive it. This way we can begin to glimpse the fundamental causes of mountain building in the Earth. I will pursue this in the final chapter of this book. Before I can do this, I need to tackle another niggling problem about the Andes; our close scrutiny of the roots to these mountains has shown that something more than just the flow of a fluid is needed to raise them. What this might be is the subject of the next chapter.

✣ ✣ ✣ ✣ ✣

Camping Out

Over the years we managed to turn the routine of camping into a fine art. The camp was arranged near our vehicle—each member of the expedition wandering off to select their own pitch for a tent. Soon the landscape would be dotted with small yellow, blue, or green domes, poking out of the low growth of tholla scrub. We would gather up armfuls of tholla branches, ideal fuel for a fire. The fire was the focus of our camp, and we spent the evening sitting around it, preparing the dinner and talking endlessly; far above, without fail, the cold night sky would be pin-pricked with bright points of light—you have to spend the night out in a high place, like the Bolivian Altiplano or great plateau of Tibet, to really appreciate just how many stars there are in our galaxy.

In the early days of our project we took it for granted that it was possible to cook outdoors only on expensive mountaineering stoves, designed to operate in the thin air at high altitude. But the jets in the burners of these stoves always seemed to get blocked; the slightest wind would either blow them out or cause the flame to stutter and lose heat. We soon gave up—it was much easier to use the campfire; with a set of nested billy cans as an oven, we even made bread and baked cakes. If the embers were buried with stones and sand when we had finished for the night, they would soon flare up again in the morning.

It was never difficult to buy fresh food in Bolivia, and sometimes we have even had it delivered directly to us in the field. In 1990 we were traveling along the eastern edge of the Sub-Andes, south of Santa Cruz, following a sandy track through the scrubby rain forest. An odd-looking horse-drawn cart approached us in the opposite direction. As the cart passed by, one of the passengers—a woman in a long dress, her head wrapped up in a large floral scarf—called out "kuchen." I was puzzled by this, but we continued on our way. About a hundred meters farther on, Leonore, who is Austrian, suddenly announced that "kuchen" meant cakes in German! We quickly reversed back toward the slow-moving cart. We had come into contact with a family from a Christian fundamentalist religious sect—the Mennonites—originally of German origin. The sect had settled here in the early part of the twentieth century, at a time when Bolivia offered a safe haven for refugees from all over the world.

The Mennonites soon established themselves as successful cattle farmers in the lowlands of Bolivia. They had, however, rejected modern dress and conveniences, still looking today like nineteenth- or even eighteenth-century pioneering settlers. The daughter of the family we had come across was dressed in a white lacey pinafore and bonnet; with her large, round face and huge eyes, she could have been from a painting by a Dutch master. Her parents spoke both German and English and were keen to talk, telling us that they were on their way to market in Santa Cruz. We bought some kuchen—rather sickly, I thought, in the humid, tropical heat—but rejected the live chickens that were offered as well. My last view of this strange group was from behind, as they moved off again, with the little girl sandwiched between her parents on the front seat, swaying from side to side as the cart lurched on the rough, dusty road.

Another time, among the high peaks of the Eastern Cordillera, we were offered some eggs by a young Indian girl who had been shyly watching us for some time as we filled a container with water from a small, clear spring. She had a dozen eggs carefully wrapped up in the folds of her shawl and was dying to make a sale. I did not particularly want any eggs—they were going to be awkward to transport—but I did not have the heart to turn her down. With great excitement she unwrapped the eggs, one by one, and I put them inside a spare hat, making a sort of nest for them in the back of the Land Cruiser among all our luggage. I soon forgot about them. That evening, as we made our way back to our base, we hit a large hole in the road rather too fast. The front of the car was thrown up into the air with a bang, and the back suspension bucked in response. I suddenly remembered the eggs. The entire nest had been hurled upward, out of the hat, crashing into the ceiling of the car—our luggage was covered with broken shells and sticky streaks of yellow yolk. The yolk managed to find its way deep into the bodywork, congealing into a hard yellow glaze. For years afterward, there were still odd patches of this glaze lurking around, summoning up memories of the Indian girl and her eggs.

Two vehicles parked next to each other in the Altiplano make an ideal arrangement for a washing line.

View from the bandstand. Aiquile's main plaza was reduced to rubble after the 1998 earthquake, and over a hundred people were killed.

Llamas watch over Bolivia's Western Cordillera, a landscape built up entirely from volcanic eruptions.

Laguna Colorada — the colored lake, home to flamingos and llamas among the volcanoes of southern Bolivia.

Our expedition in 1993 contemplates the 5,900-meter peak of Volcan Ollague, straddling the border between Chile and Bolivia. Leonore, in front, is the proud owner of a gas sample taken from one of its fumaroles, visible as a wisp of steam to the right of the summit.

Leonore (standing) and her assistant take on a boiling hot spring among the volcanoes of Bolivia's Western Cordillera.

Enveloped in choking fumes, I struggle with the gas sampling equipment on the summit of Volcan Olca, on the border between Chile and Bolivia.

Drilling for Bolivia's future. This giant oil rig penetrated nearly five kilometers of rock beneath the Altiplano, but the hole was dry.

✣ CHAPTER NINE ✣

The Subterranean Furnace

Furnace, ***n.*** Apparatus including combustion chamber in which minerals, metals, etc., may be subjected to continuous intense heat; very hot place. (*OED*)

ON JANUARY 14, 1993, around lunchtime, a small group of geologists were standing on the notoriously dangerous volcano, Galeras, in the high Andes of Colombia. The scientists were on the lip of the summit crater, a Moonlike landscape strewn with boulders and ash and stinking of sulphurous gases. They were looking for tell-tale signs of an imminent eruption. A Russian member of the group had just finished filling a glass bottle with a sample of the gas spewing out of a fumarole. A British geologist, Geoff Brown, was making some last-minute delicate adjustments to a gravimeter, trying to detect small changes in the acceleration due to gravity that might signal movements of the molten magma body deep inside the volcano. Suddenly, there was an ear-splitting crack, like the detonation of a gun. Almost immediately, blocks of rock started raining down from the air. The scientists scrambled out of the crater, running for their lives. But for most of them, it was too late. A dense cloud of superheated gas and rock, violently blasted out of the floor of the crater, engulfed them.

News of their deaths filtered back home to the geological community in Britain. I remember being overwhelmed by a feeling of shock when I first heard that Geoff Brown—a leading British vulcanologist at the Open University in Milton Keynes, not far from Oxford—had just been killed on a volcano. Only a few months earlier I had been climbing several similar volcanoes, much farther south than Galeras, in Bolivia, but still part of the same vast volcanic chain that extends along the western margin of the high Andes. The risks we had been taking were dramatically brought home to me. I am sure, however, that Geoff Brown—had he managed to make a miraculous escape—would have agreed with me that the risks were worth it in order to gain an understanding of these extraordinary features of the Earth.

We were trying to answer a deceptively simple question. Why were the volcanoes here in the first place, right above a subduction zone? Ever since the development of the theory of plate tectonics and the discovery of subduction zones in the mid-1960s, geologists had recognized a major problem with the location of these volcanoes. The problem is simply that volcanoes must lie above a source of hot, molten rock. Yet subduction zones are regions where the cold outer parts of the Earth—the tectonic plates—are dragged deep into the Earth's interior. For this reason, they would, at first sight, be expected to be places where the Earth to great depth is cooler than normal, and not the place to find molten rock. An essential step in solving this problem is determining exactly where the molten rock—the magma—comes from in the first place. And as we shall see, the attempts to do just this have led to the discovery of another mechanism, in addition to the horizontal squeezing of the crust that I have described in a previous chapter, to push up the Andean mountains and create their roots.

In the Bolivian Andes, it is clear that the volcanoes stand in line, more or less a hundred kilometers above the sinking Pacific Ocean floor. Here, the ocean floor forms part of the Nazca tectonic plate, a tongue of rock sliding into the Earth's interior at an angle of about thirty degrees in the subduction zone following the western edge of South America. A sideways view of this subduction zone immediately suggests a number of possible sources for the molten rock. It could come from the sinking plate itself. Or it could come from the intervening wedge of mantle rocks that lie between the sinking plate and the overlying crust of the Andes. Or it could come from the crust of the Andes itself, immediately beneath the volcanoes. Or, perhaps it is some combination of all the above sources? When I first started studying the Andes, all these ideas had been put forward by one geologist or another, and there was no real consensus. Whichever one was right, it had to explain the composition of the solidified magma that one can see at the surface.

Volcanoes Everywhere

You only have to look at a picture from space of the western margin of the high Andes in the Western Cordillera of northern Chile and Bolivia to see exactly what is happening here. You are looking at a landscape made entirely of solidified magma—igneous rock. From above, the high volcanic cones, built up of successive lava flows, appear almost perfectly circular, their summits highlighted by a sprinkling of snow. Between the volcanic cones are other toothpastelike flows of lava that have literally oozed out of the ground. The crust is a sieve, and molten rock has seeped out through every possible pore. And covering all the intervening spaces are sheets of pale-colored volcanic ash, blasted out of the volcanoes and now compressed into solid rock. Plumes of superheated steam rising from high up on the volcanoes, as well as numerous springs of boiling water bubbling up on their lower flanks, are clear signs that there is still unusually hot rock at relatively shallow depths.

Geologists have long classified the volcanic igneous rocks in the Andes as having what they call rather cryptically an "intermediate" composition. To understand what this means it is necessary to know something about how they have traditionally gone about dividing up or classifying igneous rock. The commonest rock of this sort is the relatively familiar basalt. Basalt erupts in large quantities from the giant volcano of Kilauea in Hawaii and Mt. Etna in Sicily, as well as underlying almost all of the sea floor, though often buried by a capping of ooze and mud that has settled out from the overlying ocean. It is a fine-grained and dark-colored, almost black, rock, composed of two distinctive minerals—feldspar and pyroxene. Another way of thinking about the composition of basalt is in terms of its chemical composition. Because oxygen is so abundant in rocks, geochemists express this in terms of the proportions of the various oxides of the chemical components, in particular silicon dioxide, which is also known as silica. It turns out that basalt is relatively poor in silica (about 50 percent by weight). Following the long-cherished and probably erroneous idea of nineteenth-century chemists, who thought that silica acted as an acid in the magma, basalt is called a basic rock (opposite of acidic) because of its low silica content. Rocks in the mantle, made up mainly of the minerals olivine and pyroxene, have an even lower silica content and are often called ultrabasic.

Another type of igneous rock, commonly found in the continents, is granite. Granite consists mainly of large, interlocking crystals of feldspar, quartz, biotite, and sometimes the dark green mineral amphibole. The crystals are a clear sign that granite has cooled slowly from the liquid state deep inside the Earth. In terms of its chemical composition, granite is rich in silica (about 70 percent by weight). Because of the high silica content it is described as acidic—in other words, it contains more of the so-called silica acid. If a magma of this composition erupts at the surface, it will cool rapidly to form fine-grained, glassy rocks called rhyolite and obsidian.

The bulk of the volcanic rocks in the Andes are intermediate between basalt and granite in terms of their chemical composition. This is why they are defined as intermediate. They are around 60 percent by weight silica, made up mainly of crystals of pyroxene, feldspar, quartz, and amphibole. The higher silica content of the intermediate lavas, compared to that in basalt, actually makes them rather sticky. For this reason, instead of the broad, shieldlike volcanoes created by runny basaltic lava flows, intermediate lavas do not flow far and tend to build up into steep-sided cones—a shape that I found from bitter experience makes the ascent of these volcanoes extremely hard work! This shape is typical of not just the volcanoes in the Andes, but those elsewhere in the world that lie above subduction zones—most famously Mount Fuji in Japan—graphically demonstrating a general association between intermediate lavas and subduction zones. But the molten rock is also very rich in gases, in particular water vapor. This makes the volcanoes particularly explosive. A volcano is really a pressure cooker, and it is the weight of the volcanic cone that holds the lid on. If the pressure builds up too much, or the lid collapses, perhaps in an earthquake-triggered landslide, the high-pressure magma can no longer be contained and explodes out of the cooker, blasting huge clouds of ash and rock fragments into the air. This ash eventually covers the landscape between the volcanic cones.

Basalt and the fine-grained version of granite, rhyolite, are also found in lesser amounts in the Andes, not just among the great chain of active volcanoes, but farther east, in the high plateau of the Altiplano. When journeying in these regions, we would occasionally pass close to small hills, a few tens of meters high, of black rock. These were mini volcanoes created by the eruption of basalt lava. They show up as black dots on pictures from space, giving the barren landscape a pock-marked appearance. Some of these volcanoes had erupted underwater in long-vanished lakes that have periodically covered the Altiplano during the last million years or so in exceptionally wet periods. The lake eruptions were very explosive, blasting out deep craters ringed by mounds of ash and broken lumps of lava. The explosions occurred when large volumes of lake water suddenly entered the volcano and rapidly chilled the hot erupting magma. One of these craters was mistakenly identified by the U.S. Air Force from satellite pictures as the impact crater of a meteorite.

So what do the compositions of the volcanic rocks tells us about where they ultimately come from? Well, the small amounts of basalt lava show that at least some of the underlying mantle is melting. This is really the only plausible way of creating basalt: the mantle rocks, made up of olivine and pyroxene, melt to form a magma from which the important components of basalt—feldspar and pyroxene—eventually crystallize out. The origin of the intermediate magmas erupting from the giant active volcanoes is more controversial. We suspected that these were mainly coming from the mantle as well, but we needed conclusive proof of this. A chance meeting at Cambridge University with Erica Griesshaber, who was a German friend of my colleague Leonore Hoke, suggested a new way of solving this problem.

Erica was completing her thesis on the hot spas that lie at regular intervals along the Rhine Valley and in the Eiffel region of Germany. The Rhine follows a rift in the crust marked by frequent small earthquakes, and the spas are fed by water that seeps out along the fault lines. The reason the water is hot is that it has risen from depths of several kilometers where the rocks are themselves a few tens of degrees hotter that at the surface. Erica had been studying the composition of the gases that are carried up with the spa waters. In particular, she was interested in helium.

The Helium Test

Helium is both a very light and an unreactive gas. For these reasons it is an ideal gas to use in lighter-than-air vehicles. Unfortunately, because helium is expensive to produce in large quantities, the early airship designers made the disastrous decision of using the cheaper but much more reactive hydrogen gas, sealing the fate of the Zeppelin airships. Helium has another property that is useful to a scientist. There are two chemically identical versions or isotopes of helium—helium-3 and helium-4, depending on whether the mass of the atom is three or four. They have different origins. Deep inside the Earth, helium-3 was forged during the early stages of the creation of the Solar System, a relict of the early Universe. On the other hand, helium-4 is a relative latecomer—a by-product, if you like, of the radioactive decay of uranium and thorium. Ernest Rutherford, in his pioneering work on radioactivity, was the first to discover this, and he called helium-4 an alpha particle—in fact, if you want to find an abundant source of helium-4, try visiting your nearest nuclear generating station. There is plenty of uranium and thorium in the Earth, and so, over time, the number of helium-4 atoms has built up. Today, they simply swamp the helium-3 in helium gas. For example, beneath your feet, in a typical piece of continental crust, there are about a hundred million more helium-4 atoms than helium-3.

It is possible to get one's hands on the helium locked up in the mantle, beneath the crust, by looking for helium gas released when parts of the mantle begin to melt. The molten rock, together with its helium, rises up to erupt on the sea floor as basalt, and these eruptions are continuously occurring along the crest of the

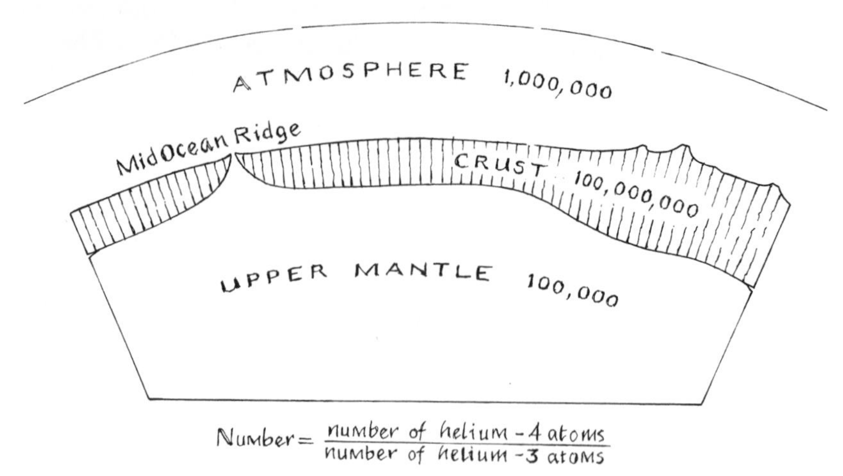

Helium gas is locked up in rocks deep within the Earth and is made up of two distinct isotopes—helium-4 and helium-3. In the crust, there are about one hundred million more helium-4 atoms than helium-3 atoms. However, the underlying mantle is much richer in helium-3, with only one hundred thousand more helium-4 atoms. These differences in composition can be used to trace the source of any helium gas coming out of the Earth. In the atmosphere, there are about a million more helium-4 atoms than helium-3.

midocean ridge, where the tectonic plates are moving apart. The minute amounts of helium trapped in small pockets of water and gas within the basalt contain about a hundred thousand more helium-4 atoms than helium-3 atoms. In other words, though the proportion of helium-4 in the mantle, beneath the midocean ridges, still far exceeds that of helium-3, there is a much higher proportion of helium-3—about a thousandfold enrichment—compared to that in the continental crust. The reason for this is simply that uranium and thorium are far less abundant in the mantle, compared to the continental crust, and so their radioactive decay has contributed much less helium-4 to the helium.

The enormous differences in the proportions of the two isotopes of helium gas, depending on whether the helium originated in the continental crust or mantle, suggest a way to fingerprint the source of helium gas coming out of the Earth. These proportions can be measured in an instrument called a mass spectrometer. If there are less than a hundred million helium-4 atoms for every helium-3 atom—say ten million, or a million, or one hundred thousand—then you can be sure that some of this helium is coming from the Earth's mantle. There is just not enough helium-3 in the crust to explain the result any other way.

Back in Cambridge, Erica Griesshaber had used the helium fingerprint to show that some of the helium gas in the hot springs of the German spa towns of the Rhine Valley was indeed coming from below the crust, in the mantle. Nobody before had suspected that the gas was coming from so deep inside the Earth. Her detective work on the helium and other gases has helped her to put together the following remarkable story of their journey to the surface. To begin with, the helium could escape from the mantle only if small pockets of the mantle were actually melting—why this is happening was a puzzle. Helium, being such a mobile gas, would immediately concentrate into the less dense magma, rising up with it toward the surface. The virtual absence of volcanic activity in the region suggests that almost all the magma never reaches the surface but is trapped deep in the crust where it begins to cool and solidify. Here, it must come into contact with deeply flowing ground water. The helium gas, along with carbon dioxide and other gases from the magma, continues its journey upward, hitching a lift with the groundwater along cracks and fault lines in the crust. Eventually, the hot water seeps out at the surface in the hot springs of the spa towns, releasing its cargo of gases.

The idea of analyzing helium emitted from volcanoes and hot springs in the Bolivian Andes first started to take root in our minds because of a common interest in making futons. Erica Griesshaber and her husband Wolfgang were enthusiastic about anything to do with Japanese culture, including sleeping on futon mattresses. It happened that both Leonore and myself had recently learned how to make a futon. When Erica found out that a home-made futon cost a mere forty pounds, compared to the hundreds of pounds that she would have been charged in a shop, she wanted to make one for herself. So we organized a futon-making weekend. And it was during this weekend that we found out much more about Erica's research, and she heard about our plans for the South American project. Erica immediately suggested that we analyze the helium coming out of the Bolivian springs and volcanoes—it will tell you immediately, she said, where the gases are coming from. Leonore's eyes lit up with excitement. She was desperately keen to come to Bolivia, but, as yet, we had not found a suitable research project that would fit easily into the Oxford Andean work. Somehow, helium and hot springs in the high Andes caught all our imaginations. Already in my mind were the experiences I had had with Raul Carrasco on my first trip to Bolivia when we visited the hot springs near Cochabamba.

Erica had perfected the technique of collecting the helium and other gases coming out of hot springs, and it seemed simple enough. Because helium is such a mobile gas, it is not easy to find a container that will hold it for any length of time. Small, helium-filled balloons that children play with at fairgrounds show this trait when they disappointingly start to deflate after a few days. Erica stored her helium samples in short lengths of special high-density copper tubing, filled by simply placing a metal funnel over the volcanic gases spewing out of the ground, or in a bubbling hot spring, and piping the gas back to the copper tube. After about fifteen minutes, sufficient gas has flowed through the system to displace all the air. At this stage, the two ends of the copper tube are squeezed off with either metal clamps or a giant pair of pliers. In this way, the malleable copper tube seals itself with a "cold-weld," forming a long-lasting, gas-tight container. The composition of the gases in the copper pipe can be analyzed later in the laboratory.

Our plan was ambitious but simple. We decided to collect helium not only from volcanoes, but wherever else it was coming out of the ground in the Bolivian Andes. This was the first time, we believed, that anybody had attempted such a large-scale study of helium in a mountain belt, covering an area of nearly half a million square kilometers. We hoped to use the proportion of helium-3 in the helium gas to fingerprint where it and, ultimately, the surrounding volcanic rocks were coming from inside the Earth.

The Ascent of Volcan Putana

The most spectacular sources of helium in the Andes are the fumaroles on the giant active volcanoes along the border between Chile and Bolivia. The volcanoes tower above the high plateau of the Altiplano, often rising to over 6,000 meters above sea level. The fumaroles are visible from miles around as white wisps of steam spewing out near summit craters, appearing deceptively close in the clear mountain air. The reality is somewhat different. Our adventures on the aptly named Putana volcano give some idea of the sheer effort involved in sampling these gases.

Putana is one of the remotest active volcanoes in Bolivia, rising up as a conical mountain on the extreme southwestern corner of the high Bolivian Altiplano. To reach it involves a long journey on sandy tracks across the barren Altiplano. The nearest gas station is in the town of Uyuni, over 500 kilometers away. We had discovered a fumarole on Putana by chance while trying to reach a hot spring in the same region. Far up on the flanks of the volcano we could see, with our binoculars, the tell-tale white cloud of the fumarole. Getting to it seemed virtually impossible, involving the scaling of near vertical cliffs of black volcanic rock. So we abandoned the idea of sampling the helium in this fumarole and continued on to a well-known hot spring near a sulphur mine called Rio Blanco.

We skirted a large, dormant volcano, following a track through deep, sandy volcanic ash. The buildings of Rio Blanco mine formed a long, low, mud-built block, roofed with corrugated iron. There seemed little activity here, an impression that was confirmed by the presence of a rusty, Russian-built tractor, parked outside with its engine partly dismantled. It turned out that the mine was deserted, except for two caretakers and their families. They were pleased to see us, having been virtually abandoned by the mine owners some months before. They were particularly interested in our car battery. Apparently, they had been desperately trying to call up the mine owners on the radio for more food, but their own battery had failed. We disconnected our battery and carried it into the main mine office.

The radio soon crackled into life as one of caretakers intoned the call sign of the mine, trying to reach another mine several hundred kilometers away.

The hot spring was only a few hundred meters from the mine building. The miners had built a hut over it, using the hot spring as a sort of bath house. That evening, after Leonore had collected some of the gas from the spring in a copper tube, we set up camp in one of the many empty rooms of the mine building. We asked the caretakers if they knew of any other hot springs or volcanic fumaroles in the area. One of them told us about a fumarole on Volcan Putana—this could have been the one we had spotted previously. He said that we could reach it in a day from the mine. He had, in fact, guided a group of geologists up there before. For a fee he would show us the way up the following day. And so it was that we began our ascent of Volcan Putana at about six o'clock the next morning. Our group consisted of five members of the Oxford project—Leonore, Lorcan, and myself, together with Susan and Jurgen, who had both joined us for a couple of months—and our guide. We could drive the first part of the way, and our route took us past the mine quarry itself. Our Toyota Land Cruiser lacked power in the freezing cold dawn air as we struggled up the steep track to the quarry. Here, when the mine was working, pure, bright yellow sulphur was dug out of the side of the mountain.

We left the vehicle at an altitude of about 5,000 meters and scrambled on foot up a steep scree slope of volcanic ash. I monitored our climb with a small precision altimeter. The volcano summit was just under 6,000 meters above sea level, so I reckoned that we had a kilometer climb ahead of us—a hard climb, but easily possible in a day. To begin with we climbed steadily, ascending nearly 600 meters. Progress was slow going, following the principle of three steps forward and two steps backward in the scree. At altitudes over 5,000 meters, the effort is approximately twice that at sea level to achieve the same result. Sometimes it seems more like ten times as much. This is because there is only half as much oxygen, so each breath has a much smaller effect on recharging the blood. The subzero dry air burns the throat and lungs, making each gasp for oxygen very painful. But the climbing is not difficult from a technical point of view—it merely requires dogged perseverance and determination.

At 5,600 meters I began to realize that something was wrong. Our guide seemed unsure of himself, dodging this way and that on the slope searching for something—he looked lost. I tried to reassure myself that this was not the case by asking him what he had done when he guided the other party. It turned out that he could not remember because this was over five years ago. Suddenly he disappeared over a ridge. When we got to the ridge we saw him scurrying down a steep slope, descending rapidly with great leaps in the volcanic scree. We were not on Volcan Putana after all, but another unnamed volcano that stood between Rio Blanco and Putana. Ahead of me, on the other side of a sandy plain, stood the near perfect cone of Putana, taunting me with its height. We would have to descend the volcano we were on and start climbing Putana from an even lower elevation than where we had left our vehicle. My heart sank and my energy seemed to drain away.

At that point, Susan decided that she had had enough and wanted to go back to the mine. I had visions of the party getting dispersed and lost in the volcanic landscape, where one hill looked much like another. So, summoning all my breath, I shouted to Lorcan to go back with Susan. Lorcan, naturally enough, was not very happy about this. However, I was too worried about Susan being on her own to care. Reluctantly Lorcan caught up with Susan, and I watched them disappear back over the ridge crest on their way home. Leonore, Jurgen, myself, and our guide continued on. The climb, instead of being a single thousand-meter ascent, was turning into about 1,800 meters of ascent, and the same amount of descent, up and down the flanks of two volcanoes; all this had to be repeated in reverse on the way back.

The final few hundred meters of the climb were the most exhausting of my life. By now my world had focused down to the next ten paces. We had at last reached the snow line, and with each step I sank calf deep into the soft snow. I willed myself on by selecting a rock a few meters ahead of me, pulling my legs out of the snow and counting my steps. Sometimes I did not even have the energy to reach the rock but would collapse in the snow on my back, panting heavily. In my desperation to finish the climb I was ascending too fast, squandering my remaining resources of energy. I could see the others far below me. I had to slow down, but I was now consumed by a new worry. Would we manage to get off the volcano before it got dark? We had been misled by our guide and had planned for only a one-day climb. I was not sure if we would make it through a night at such a high altitude without a tent, sleeping bag, or hot food. The night air temperature would plummet to −20°C or −30°C. Yet I was loath to give up, having come so far.

Gradually, the sides of the volcano began to flatten out. Ahead, I could make out fingers of white and yellow cloud streaming up into the deep blue sky. However, the intensity of my concentration must have been such that I do not remember the view from this high place—stretching far into the distance, southern Bolivia must have been laid out as a vast panorama of cinder and lava, heaped up into dark high volcanic peaks or spread out as grey carpets of ash. I do remember, though, the harsh smell of sulphuric acid as I stumbled toward the main active crater. When I finally stood on the edge, looking down at a fuming hell of steaming sulphuric acid and sulphur, I realized that it would be impossible for us to sample any of the gases. One side of the crater had been blasted away in a previous volcanic eruption, and the gases were streaming out of a sheer cliff face of volcanic rock. I looked around desperately for some part of the crater we could reach. And then I saw the fumarole we had first spotted with our binoculars from far below, forming a single column of white steam spewing out of a small side peak of the volcano, to one side of the crater. We would be able to get to this, but it would mean descending another hundred meters or so on the far side of the volcano. When Leonore and Jurgen finally caught up, I pointed this out to them and we wearily made our way down yet another scree slope to the fumarole.

It proved to be very difficult to sample the gas. We were choking in the fumes and found it hard to channel the steam into our copper tube. As we fumbled with the equipment, time was ticking on. It was now late afternoon, and our guide was getting very impatient. He also was worried about the impending night. Unlike us, who at least had windproof warm clothing, he was wearing a shirt and sweater with thin cotton trousers, and sandals on his feet. Finally, at four o'clock we were back on the main summit and ready to leave. Our guide needed no encouragement but shot off down the scree slopes and was soon far ahead of us. Leonore and Jurgen were much slower, and so I was left pig in the middle, trying to act as a link between the guide and the others. I knew that if we lost sight of the guide we would be in trouble, because with all the ascents and descents our route had taken on the way out, I was unsure of the way back to the vehicle. With night coming on, we could not afford to get lost.

We finally reached the bottom of Putana. The guide was far ahead, striding across the sandy plain to the next volcano. I virtually had to run to keep up. As I started the exhausting ascent of this second volcano, I could just make out Leonore and Jurgen in the twilight still on Putana. I pleaded with the guide to slow down, but he was unmoved. Getting back soon was becoming a matter of life and death. Our route angled around the second volcano, making for a pass on a shoulder ridge. There came a point where I could not keep following the guide and keep the others in view as well. In the gloom they might not see where to go and miss the pass into the Rio Blanco Valley. I decided to stay put, keeping an eye on the guide who was silhouetted against the sky while flashing my flashlight at Leonore and Jurgen. This way I could signal to them, showing them what to head for. To my relief, they understood and signaled back. I then continued on up to the ridge, forcing the guide to wait with me. From there, I signaled continuously with my flashlight to mark the way up.

I think the guide at last began to realize that we were in danger of losing the others, and so he crouched beside a rock, sheltering from the wind and warming himself with his arms. I was alarmed to find that my flashlight batteries were failing. This was probably because the air temperature was now well below zero. I flashed every few minutes for about forty-five minutes, willing the others to flash back. By now it was fully dark—there was no moon, and the rocky mountainside was pitch black, vaguely outlined by the slightly lighter night sky. I was beginning to panic because there was still no reply from Leonore and Jurgen. I tried shouting, but I did not have much breath and my voice sounded thin in the cold air. Should I go back and look for them at the corner, where the trail angled around? And then, to my relief, I saw a single flash of light. I flashed back, and my flashlight failed almost immediately afterward. My response, however, was enough. We all reached our vehicle safely at about ten o'clock, nearly sixteen hours after we had left it that morning. Lorcan and Susan had already returned to the main mine buildings on foot, and they were now anxiously waiting with the other Bolivian miners for our return. We were very tired and cold, and I had a splitting headache. But we had got our helium sample.

Aguas Calientes

Hot springs do not just occur close to the active volcanoes; there are hot springs all over Bolivia, often in the most unexpected places. They provide the inhabitants in the remote, mountainous parts of Bolivia with a free and almost limitless supply of hot water. So if you ask in a village whether there any hot springs (*aguas calientes*) in the area, somebody will usually be able to tell you. This is the way Leonore found many of her sampling sites. If a spring was in the bed of a river, the first sign of it was often an abundant growth of algae, adding a splash of green to the brown landscape. In the evening or early morning, a thin mist would hang over the surface of the slightly warmed water.

Sometimes the hot water fountains out under pressure rather like a garden sprinkler, catching the sunlight and turning it into a beautiful arc of a rainbow. Most of the hot springs are only warm, with the water temperature between 30°C and 40°C, seeping out to form a small, dark liquid pool among the rocks. Usually, the only sign of any activity is the steady stream of carbon dioxide bubbles gently stirring up the sandy bottom. Leonore would capture these springs in much the same way that one might net a rare insect, placing a funnel mounted on a long stick over a particularly vigorous stream of bubbles. One could follow the bubbles as they made their progress along the coils of tubing into the copper pipe.

Some hot springs are virtually boiling—their surfaces vigorously heaving and splashing. Once we camped close to one of these springs. The noise of the spring kept me awake, but the steam provided a small pocket of warm air, warding off the extreme cold of the Andean night. That night the boiling water was also slowly cooking a meal. This new culinary opportunity brought on a rash of ideas. We placed a pot of rice, some potatoes, and eggs in the boiling water, wedging them carefully with rocks to keep them in position. To our delight, the next morning we found the food ready to eat. A new use, we thought, for a Bolivian hot spring, and a departure from the usual sampling for helium. Leonore, though, would never abandon a hot spring without a helium sample.

A most intriguing spring was what we called a cold bubbler. The water is cold but full of gas bubbling out with a distinct glug-glug sound. We found one of these in the high plateau of the Altiplano, looking rather like a giant termite hill. It was clear that a limestone scale was precipitating out of the spring water. Over time, this scale had built up, so that now the water was bubbling out nearly three meters above ground level. Several other mounds nearby were dry and seemed to be the remains of extinct cold bubblers.

Berlin—Analyzing the Samples

By the end of our first major field season, Leonore had acquired a neat stack of sealed copper tubes, full of gas from Bolivian hot springs and fumaroles. And the contents of each copper tube had a story behind it. We now needed to analyze the samples. Unfortunately, the department in Oxford did not have the facilities to measure isotopes of helium. Erica Griesshaber, who had helped us set up our helium project, suggested that we use a laboratory in Berlin where another of her supervisor's students, David Hilton, was working. And to our surprise we found that David was also studying helium from the Andes. The discovery that somebody else is working on the same thing is always a scientist's worst nightmare. However, to Leonore's enormous relief, David Hilton had been working in Chile and Argentina, far to the south of where we had been sampling. And there was a dividend—he was keen to work with us and help us analyze our samples. We must come to Berlin.

It turned out that in the autumn of 1990, just after the Berlin Wall had come down, I was invited to give a talk on my Andean research to a group of geophysicists at the Frei Universitat in Berlin. Leonore decided to come with me to start work on the helium samples. We found David Hilton's laboratory—also part of the Frei Universitat—in an old Berlin house close to the water tower where Heisenberg had tried to build the world's first nuclear reactor. The laboratory was taken up with several large mass spectrometers and the spaghetti-like system of pipes and pumps needed to handle the gases. The whole system operates at high vacuum, and the laboratory sounds like a giant refrigerator, with a steady background gurgling noise from the pumps.

Leonore was quickly introduced to the analytical procedure. First, the gas sample in the copper tube must be transferred to special glass vials that can be inserted into the mass spectrometer—this transfer is done via a network of glass tubes called an extraction line. This turned out to be more difficult than it sounds. The main problem was devising a gas-tight connection between the copper tubes, battered after months of traveling around Bolivia, and the extraction line. Before the sample is released into the system, the connection is tested by pumping out the extraction line. The vacuum test always adds an element of tension to the analysis. If it is not possible to reach a high vacuum, even with the pumps running for several hours, then it means there is a leak. And a leak can take hours or days to fix and could even result in the loss of the sample itself. If you have just scaled a volcano or traveled to the remote corners of the Bolivian Andes to collect a sample, its loss can be heartbreaking. Eventually Leonore devised a simple connection, using an adaptor and silicone O rings, that was gas tight.

Once a vacuum tight connection between the copper tube and extraction line has been made, the vacuum pumps are disconnected and the gas sample is released into the evacuated system. This is done by slightly loosening one of the clamps that hold the cold weld seals at the ends of the copper tube. Even with only a very slight easing up of the cold weld, the high vacuum in the extraction line will immediately suck the gas out of the copper tube. This gas will be a mixture of water vapor, carbon dioxide, helium, and many other gases. The next stage is to concentrate the helium by separating out the abundant water and carbon dioxide. These are called condensable gases because they can be liquefied at only moderately cold temperatures—compared to helium, which is only liquid close to absolute zero. The gases pass through glass bottles bathed in water ice or dry ice—cold fingers in the jargon of chemists. The condensable gases are frozen out of the gas sample in the cold fingers and can be analyzed separately. The rest of the sample passes through various inert scrubbers or filters designed to absorb any other unwanted gas. The final helium sample is trapped in a specially designed glass tube. This tube is sealed off and disconnected from the extraction line by pinching off the glass with a blow-torch. The vial of helium gas collected this way—called a break-seal—is ready to be put into the mass spectrometer.

The mass spectrometer is one of the most powerful tools in the armory of a geochemist. It is an ingenious machine, designed to separate out atoms with small differences in their atomic weights. For example, mass spectrometers can easily separate out isotopes of elements, such as lead, where there is only a 1 percent difference in their weights. With the isotopes of helium, the difference in weight is relatively large—helium-4 is a third heavier than helium-3. In essence, the machine consists of a bent metal tube that passes between the poles of a giant magnet. Before any measurements are made, the inside of the tube is evacuated.

The sample is let into one end of the tube and is heated to a high temperature; this has the effect of giving the helium atoms an electric charge. The charged atoms accelerate toward an oppositely charged grid, positioned a little way along the length of the tube. Many of the atoms pass through the holes in the grid at high speed, on their way farther down the tube. They stream around the bend between the magnet, finally reaching a detector at the far end. One can think of the detector as an archery target. It is possible to adjust the strength of the magnet, so that only atoms of a particular mass end up in the bull's-eye—all other masses will end up somewhere on the edges of the target. And it is only those atoms that end up in the bull's-eye that will be counted. By pulsing the strength of the magnet, so that the two masses arrive alternately at the bull's-eye of the detector, it is possible to measure the proportion of two isotopes with different masses. In this way, the atoms of both masses can be counted simultaneously.

Helium Results at Last

David Hilton had agreed to analyze the first batch of samples himself in the mass spectrometer. He thought that we would get the results quicker this way as the Berlin mass spectrometer was a bit tricky to use. So, back in Oxford, we waited for the results to be sent to us. One day, Leonore came to my office with a fax from Hilton in her hand. A long list of numbers on the fax contained all one would ever want to know about the helium isotopes in our gas samples from Bolivia. We scanned the list trying to come to grips with what the numbers meant. The proportions of the two isotopes of helium—helium-3 and -4—varied considerably between samples. At first glance, the results looked interesting.

We could use the raw measurements of the ratio of the two helium isotopes in any gas sample to work out what proportion of the helium gas originally came from the mantle, and what proportion came from the crust. This is because each source—the crust and mantle—has a characteristic proportion of helium-3 and -4. Unfortunately, the samples were labeled with code numbers. It would not be until we had plotted the sample sites with their helium composition on a map that we would be able to make sense of the results. Scientists search for a pattern in scientific observations. It is often the pattern rather than the individual observations that can be so convincing. And, as we plotted the data, it became clear that there was a simple pattern to the proportion of helium isotopes in our Bolivian samples. Over the years, as we added more samples, expanding the work and making more and more analyses, the pattern became clearer.

The pattern is easiest to appreciate if one imagines crossing the Andes from the Pacific coast to the Amazon jungle. Near the Pacific coast, the proportion of helium-3 in gas from hot springs is extremely low and virtually all of the helium is helium-4, scavenged from the Andean continental crust by deeply flowing groundwaters. This helium is being created as the high concentrations of uranium and thorium in the crust undergo radioactive decay, emitting helium-4 particles. However, as one approaches the active volcanoes of the Western Cordillera, the first sign, or sniff if you like, of helium from the mantle appears. The gas becomes richer and richer in helium-3, and one can easily calculate that over 80 percent of the helium coming out of the volcanoes themselves, or from hot springs on their flanks, must be coming from the mantle, deep beneath the Andean crust. Here was clear evidence, at last, that the mantle was the main source of the molten rock erupting in these volcanoes, releasing its helium at the same time.

We were amazed to discover that a significant amount of the helium gas coming out of hot springs in the Altiplano and western margin of the Eastern Cordillera, up to 350 kilometers east of the great chain of volcanoes, was also coming from the mantle. You would have to go over 400 kilometers east of the volcanoes before you would observe helium again coming solely from the Andean crust. The picture we were getting of the high Andes through "helium spectacles" was a surprising one, suggesting melting of the mantle in a vast region, both beneath the chain of active volcanoes and much farther east as well, and more extensive than anybody had suspected before. We now needed to find out why the mantle was melting in the first place.

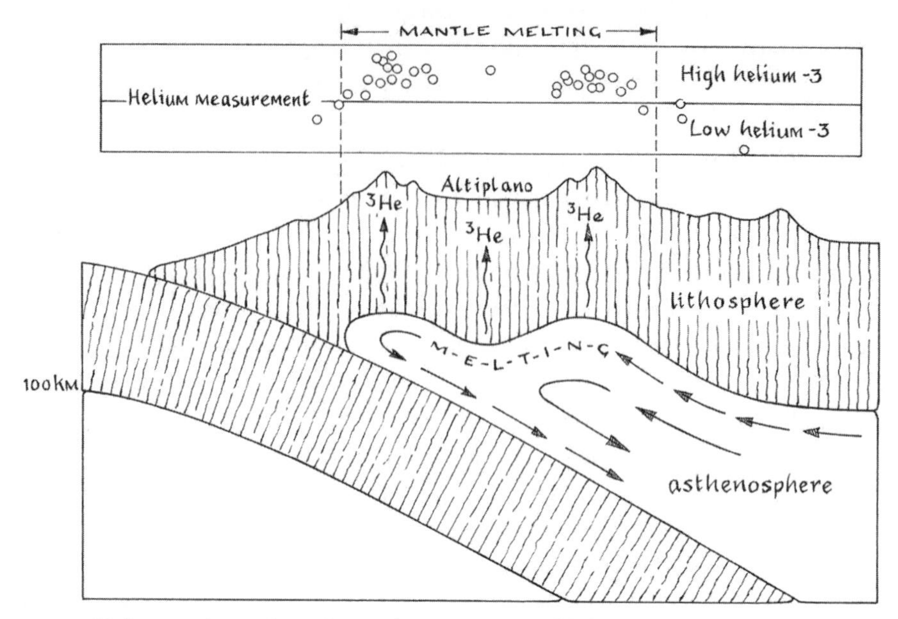

Helium and mantle melting: the proportion of helium-3 in gas leaking out of the crust shows that the mantle beneath the high Andes is melting. The molten rock rises up to erupt in volcanoes on the surface releasing its cargo of helium-3. The mantle itself is flowing, dragged around in a sort of eddy current by the sinking oceanic plate.

Making Volcanoes with Water

The common thread that runs through all the many aspects of the volcanoes of the high Andes turns out to be water. Leonore found that many of her gas samples contained a high proportion of water vapor, in addition to many other gases including carbon dioxide, nitrogen, and argon—helium only makes up a very small fraction of the sample. The molten rock at depth is rich in this water. This is one of the reasons that these volcanoes are so explosive—it is the high pressure of the superheated water vapor, combined with the sticky nature of the lava, that makes the volcanoes into such an unstable pressure cooker, always on the verge of blowing up. And, as we shall see, water may be the reason that the rocks are melting in the first place. Therefore, it is natural to ask where this water is coming from. The answer seems to lie in the isotopes of some of the other elements in Leonore's gas samples, which, like helium, have a story to tell.

Water is, of course, hydrogen and oxygen. Hydrogen has two isotopes. The heavy isotope of hydrogen, called deuterium, has an atomic mass of two, compared to one for ordinary hydrogen. Water enriched in deuterium is sometimes referred to as heavy water—this is an important moderator of nuclear reactions, used in certain reactors. Natural samples of water contain a small proportion of heavy water, but the exact proportion depends on where the water comes from. For example, rain water is enriched in light hydrogen, while ocean water is relatively enriched in deuterium.

Leonore found that when she measured the ratio of light to heavy hydrogen in the water samples from the volcanoes, using much the same techniques as those for measuring helium isotopes, the ratio came out to be very similar to that of ocean water. Could it be that some of this water has come directly from the ocean? An analysis of the nitrogen provides support for this idea. Leonore found that the proportion of nitrogen gas in the samples, compared with other gases such as argon, was unlike that of nitrogen and argon in the atmosphere. This feature of nitrogen has been found in gases coming not just from the volcanoes in Bolivia, but from many others in the Ring of Fire around the Pacific margin. This strongly suggests that it is a feature related to the subduction process itself. One possible source of the "extra" nitrogen is the cooked-up remains of dead organisms on the sea floor that are known to be rich in nitrogen.

All this begins to make sense if one considers the overall shape of the subduction zone. It is clear that some of the closest sea floor lies directly beneath the volcanoes, forming the top part of the sinking Nazca plate as it slides underneath South America. During the last ten years, geophysicists and geochemists have attempted to work out the chain of events that might link this sinking slab with the eruption of molten rock from the overlying volcanoes. Their ideas have been largely based on the results of experiments with molten rocks in a variety of conditions, put together with the recent discoveries about the nature of the sea floor. Our new results from the Andes provide yet more evidence that these ideas, outlined below, are basically right.

Swan Song of an Ocean

The crust beneath the deep ocean floor along the western margin of South America has subtly changed since it was created at the midocean ridges. This is partly because there has been pervasive passage of water through the ocean floor rocks. Cold ocean water percolates down through innumerable cracks. As it sinks it encounters hotter and hotter rocks and eventually heats up itself. The hot water is now too buoyant to sink farther but instead starts to rise up, reaching the surface at deep-sea black smokers and hydrothermal vents. But the deeply penetrating water has had sufficient time to react strongly with the rock, changing the composition of ocean crust by adding water and salts, and scavenging elements such as iron and manganese. In addition, as the ocean floor moves away from the midocean ridge, a veneer of deep ocean sediment is deposited on top, forming a new geological layer.

In the ocean trench, about a hundred kilometers off the coast of Peru and Chile, the altered ocean floor, together with part of its covering layer of sediment, forms part of a slab of rock that sinks back into the mantle, embarking on a long journey into the planet's interior. Here, it begins to feel the effects of the high temperatures and intense pressures inside the Earth. As it continues downward, any water locked up in the rocks is gradually squeezed out. For instance, when the slab reaches a depth of 50 kilometers, it will be at a temperature of hundreds of degrees centigrade, subjected to a massive pressure fifteen thousand times that on the Earth's surface. In response, changes occur in the rocks. Water and carbon dioxide are driven off, and new minerals, such as amphibole and garnet, which are stable at the higher temperature and pressure conditions, start to form. The intense shearing caused by the sliding action of the sinking slab may further raise its temperature. Eventually, parts of the ocean crust may become hot enough to melt, releasing more water vapor. At depths in excess of a hundred kilometers, minerals like amphibole collapse under the immense pressures. Even more water, which was locked up in the amphibole mineral, is squeezed out.

All this water seeps out and rises up into the surrounding hot mantle; this meeting of water and hot mantle is, in some ways, the moment of conception of both the overlying volcanoes and, as we shall see, the crust itself. Something extraordinary happens: the water triggers a small amount of melting in the mantle, in much the same way that salt—or even alcohol—when sprinkled on a frozen road can destabilize the ice and turn it into slush. In the Earth's mantle, the molten rock begins to pool and, being less dense than the surrounding rock, continues to rise up as a body of magma into the overlying crust of the Andes, taking with it a cargo of gases including water vapor, carbon dioxide, and helium.

Geologists have a good understanding of what happens next, based on long study of the minerals that make up igneous rocks combined with experiments trying to melt them. A small portion of the new magma will rise straight up to the surface and erupt as basaltic lava flows, revealing the composition of the original molten rock. But for the bulk of the magma, the upward journey is more protracted and involves cooling and crystallization of new minerals deep below the surface. However, the full range of different minerals that will be present in solidified rock do not crystallize out all at once; instead, the minerals crystallize in a particular order. The first crystals to form are those that have the highest melting points—these are the silica-poor but magnesium- and iron-rich and relatively dense minerals, olivine and pyroxene. They tend to sink to the bottom of the body of magma, leaving behind a liquid that is relatively poor in magnesium and iron but, importantly for the generation of intermediate igneous rocks, enriched in silica. Thus, the magma undergoes a sort of refining process, which progressively increases the concentration of silica in the remaining liquid. Finally, at lower temperatures, minerals typical of intermediate and acidic rocks, such as quartz, alkali feldspar and mica, start to crystallize.

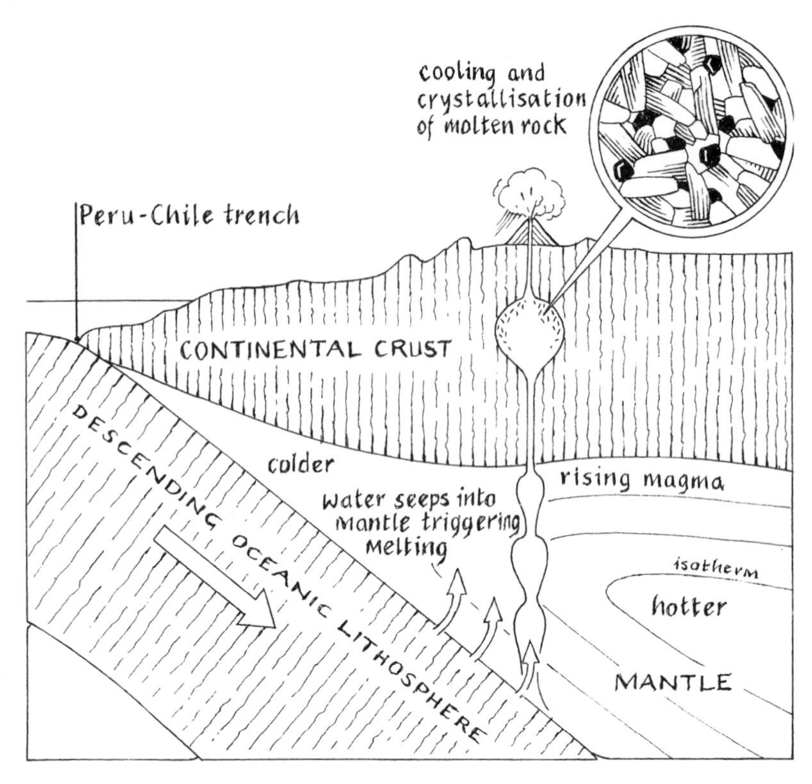

Molten rock erupting in an Andean volcano has completed a remarkable journey through the Earth. The melting is triggered at depths of about 120 kilometers by water seeping up from the descending oceanic lithosphere. The resultant magma pools and rises up through the crust, slowly cooling and crystallizing, before finally reaching the surface with the characteristic andesitic composition of the great chains of volcanoes that encircle the Pacific Ocean.

Building the Crust

I have now come a long way in explaining how volcanic rocks with an intermediate or even acidic composition end up erupting from the volcanoes. In essence, the explanation is that the original primary basic (basaltlike) composition of the magma in the mantle separates into two parts. One part is very poor in silica and has an ultrabasic composition, containing the olivine and pyroxene crystals. The other part forms a residue relatively enriched in silica. The ultrabasic portion, being both relatively dense and the first to crystallize out from the melt, is mainly filtered out at the base of the continental crust. Because it has essentially the same composition as mantle rocks, it merely becomes part of the mantle. The remaining magma continues on up into the continental crust, undergoing its refining process as it cools and crystallizes, ending up with an intermediate or acidic composition. Sometimes, fragments of sedimentary rock, looking a bit like blobs of melting ice cream, show that a portion of the surrounding crust gets mixed up with and contaminates the magma, further adding to the variety of its composition.

Simple considerations of the steady cooling of the magma beneath the volcanoes suggest that most of it must solidify deep in the crust and never reach the surface; some geologists have calculated that this could be more than two-thirds, and perhaps as much as nine-tenths, of the original volume of molten rock. The volcanic activity at the surface is really just the tip of a volcanic "iceberg." Most of the magma lies trapped deep beneath the volcano, forming vast bodies of granite and more intermediate rock. And over time, this molten rock will build up the thickness of the crust. In the Andes, the volcanic activity at the surface suggests that a layer of new igneous rock, at least half a kilometer thick, is added to the crust beneath the volcanoes every million years. And looking back over the last few tens of millions of years, since the Bolivian Andes first started to rise, the same crust must have thickened up at least twelve kilometers simply due to the melting of the underlying mantle. Our new helium measurements now show that this melting is taking place over a much wider region, extending as far as the Eastern Cordillera. All this will also have added to the thickness of the crust, but to a much lesser extent than beneath the great chain of volcanoes in the Western Cordillera, where the surface volcanic activity has clearly been far more prolific.

We are now, at last, ready to tackle the problem that emerged when we considered mechanisms for the thickening of the crust beneath the Andes (see chapter 6). Measurements of the folds and faults in the rock layers—the products of horizontal squeezing of the crust—showed that there has been enough squeezing during the life of the Andean ranges to build most, but not quite all, of these mountains and their deep roots. It is clear, now, that the remaining volume of the roots must have been topped up by the magma coming from the mantle, on its one-way journey into the Andean crust. In other words, the Andes have been raised by two distinct mechanisms. In the east, beneath the Eastern Cordillera, the crust has been predominantly thickened by squeezing, making the mountains and their roots. Farther west, however, beneath the volcanoes of the Western Cordillera, both squeezing of the existing crust and the addition of new molten rock from the mantle have been a significant cause of this thickening. In fact, we are witnessing here something even more fundamental—the creation or growth of the continents themselves. Over geological time, the molten rock, not just in the Andes, but above subduction zones all over the planet, has built up the continental crust. And the helium coming out of the high Andes shows that this is still going on today. Whenever I journey among the great volcanoes of the Andes I still find it an amazing thought that beneath my feet a fundamental process is going on that is creating more of the dry land on which we live.

All the work among the volcanoes has produced vital new pieces of information in the bulging geological dossier of the Bolivian Andes. This dossier documents our progress in understanding the growth of these mountains; we have a good idea of what has happened, and we also understand many of the fundamental processes that have controlled these events. But, stepping back from our work, it is clear that all this fits into a much bigger picture; mountain ranges are both an expression and an integral part of the workings of the Earth as a whole. This is the subject of the final chapter in my narrative.

✣ ✣ ✣ ✣ ✣

Health at High Altitude

Despite the battery of injections at the Oxford University medical service, including a giant one for hepatitus that had to be given with a horse syringe, we could never escape the hazard of high altitude. This problem becomes significant above about 3,000 meters, the altitude of most of the Andes where we were working. At these altitudes the concentration of oxygen in the atmosphere is just too low for the body to function normally without some major physiological adjustment. In fact, it always amazes me how adept the body is at adapting to these rarified conditions, especially as most mammals, including ourselves, have evolved close to sea level. Much like the human body's ready adjustment to the condition of weightlessness in space, it must be a fortuitous by-product of another aspect of our physiology.

A typical sign of mild altitude sickness is a blue-red hue to the face and lips, accompanied by a splitting headache. Unfortunately, painkillers only hinder the acclimatization process, so it is best just to try to weather this storm of pain. At the same time, the slightest exertion, such as walking slightly uphill or climbing stairs—something one always seem to be doing with heavy packs on arrival in La Paz—seems particularly exhausting. One tends to lose one's appetite and have enormous difficulty sleeping. I have found from bitter experience that it is almost impossible to sleep after a heavy evening meal—you spend the night thrashing around while your stomach fights with the brain for the available oxygen supply.

Despite the discomforts, your body is working hard to come to terms with the thin air, as it manufactures extra red blood cells in a frenzy of activity. This is the main process of acclimatization, and most of it has taken place within two days of your arrival at high altitude. Thereafter, there is a much slower physiological adjustment over several weeks. Luckily, we have never had any bad case of altitude sickness on our many expeditions. The nearest I came to this was when I started climbing the high peaks of the Quimsa Cruz range, southeast of La Paz, within a few days of arriving in Bolivia. I was trying to collect samples of granite for dating. We were working during the day at altitudes well over 5,000 meters, and our campsite was not much lower at 4,900 meters. When I came back to my tent on the second day I noticed that my hands were tingling. The tingling sensation became stronger and spread to my legs and feet. Soon I found it difficult to hold an object or walk. It is often said that there are only three ways to deal with altitude sickness: descent, descent, and descent. Sometimes, trying to descend can make matters worse if it involves, as is often the case, crossing an even higher mountain pass. We managed to drop down to about 4,000 meters—still a very high altitude, but not so extreme as our original campsite. After a few days the tingling sensations began to fade and I was able to ascend again.

✣ CHAPTER TEN ✣

Putting Up Barriers

Barrier, ***n.*** Any obstacle, boundary or agency that keeps apart.
(*OED*)

I REMEMBER ONCE hovering in a helicopter high above the narrow ranges of the Southern Alps in New Zealand. It was early in the morning, and the air above the mountains was crystal clear, tinged only by the rays of the rising Sun. Looking north, I could see in front of me jagged peaks of dark rock, with ice and snow clinging to their sides. On my right, far to the east, lay the McKenzie Country—an arid landscape of rolling hills in the foothills of the Alps. But on my left, in the west, the ground was obscured by a continuous blanket of white clouds. The mountains stood like a wall between these two regions, with the clouds lapping at the steep western front of the Southern Alps like water on a lake shore (though in a few places white, feathery fingers of mist had crept through valleys, reaching out for those dry and clear lowlands in the east). This aerial perspective clearly revealed the enormous impact of high mountains on the weather. The Southern Alps stand as a nearly impassable barrier to the moisture-laden air in the west, blown off the Tasman Sea. This air releases its cargo of water on the western flanks of the mountains, making this region one of the wettest places on Earth, cloaked in dense, temperate rain forest. But the eastern side of the Alps, sheltered from the rain, remains parched and brown.

I could have had a very similar view of the weather near mountain ranges in many other parts of the world. The great ranges of the Himalayas trap the moisture in the air from the Indian Ocean, turning the great plateau of Tibet into a freezing desert. And the Andes themselves form a barrier between the humid rain forests of the Amazon and the dry Atacama Desert. Mountains, however, have a much more profound effect on the weather than just blocking the flow of moist air. They can also be the main driving force for these winds. This was forcibly brought home to me when I camped on the western margin of the Bolivian Altiplano beneath the high conical volcanic peaks of the Western Cordillera. Each day we experienced an almost identical weather pattern.

The days were cloudless and hot, and the nights were incredibly cold. Each morning when I awoke, I had to wait for the shadows of the volcanoes to retreat from our campsite. Only then was it bearable to leave the warmth of my down sleeping bag. The morning air was absolutely still. But without fail, at around ten o'clock, a violent wind blew up from the west, and our campsite was immediately turned into a raging dust storm. The wind blew until the early evening, when it dropped as suddenly as it had come. The reason for this turned out to be quite simple. Our campsite in the high Altiplano was at an elevation of about 4,000 meters above sea level. During the night the air temperature was well below zero. However, in the morning, the Sun's rays quickly heated up the brown, rocky landscape, warming the air in the process. The warm air began to rise, eventually sucking in cold air from out to sea and creating the turbulent winds that had made the day so unbearable for us. When evening came, the Altiplano quickly cooled down again and the winds vanished.

Meteorologists agree that mountains can influence not only the local weather, but the entire climate of the planet. This way they can control the lives of living organisms, including ourselves. And the record in the rocks — for example, the evidence from the Andes themselves — suggests that they may have made the Earth a slightly more or slightly less comfortable place for life in the past, as mountain ranges have come and gone in our planet's long history. This brings us to the fundamental idea that the surface of the Earth is inherently unstable, and this instability can have far-reaching effects. Mountains can rise and fall, ebb and flow, having their brief moments of glory and power when they tower above the surface of the continents.

So what forces in the Earth ultimately control this motion? This is essentially the same question I posed at the beginning of this book. Unfortunately, not all of the answer lies within our present clear scientific vision but is around the corner of scientific research. In trying to peer around this corner, I have had to stick my neck out farther than I have done so far in this book, in some places entering the realm of speculation, though there are some good signposts to indicate the way. It is certainly worth the effort because the answer will not only help to pull together the many strands of our investigation in the Andes, described in the preceding chapters of this book, but will also reveal the wider role of mountains on Earth.

A Saucepan of Hot Syrup

If you lift an object, you are working against gravity. You therefore have to expend some effort, supplying a force that will cause the object to rise. Exactly the same principle applies to mountains. Mountains are raised up because of some force within the Earth that counteracts gravity. In fact, such a force has also to work against the inherent strength or stiffness of the rocks themselves, in the same way that stirring sticky syrup, or kneading bread, is hard work, even out in space where the force of gravity is tiny. Ask any Russian astronaut who has worked on the MIR space station. So where does the force come from to push up mountains? Curiously enough, the answer is that it comes from gravity itself. This may seem to be a direct contradiction of the very principle I have just described. But consider the workings of an old-fashioned clock. The clock consists of a large number of interconnected cogs that move the hands and ring the chimes. The whole mechanism is driven by the gravitational force of a suspended heavy weight. As the weight falls, it pulls on a cable, turning a drum that drives the clock. And the operation of the clock involves lifting parts of its mechanism.

We need to find a clocklike mechanism in the Earth. In fact, the mechanism is staring us in the face. It is the large-scale motion of the tectonic plates, created at the midocean ridge, then gliding over the surface of the Earth before sinking back into the interior again at a subduction zone. The plates are just the top of a vast, churning cauldron of flowing rock within the Earth. This has been one of the great revelations about our planet in the last forty years.

The Earth is behaving very much like a saucepan of hot syrup—molasses or treacle—subjected to the force of gravity and heated from below. The syrup at the bottom of the saucepan is much hotter than that near the surface. For that reason it is also very slightly less dense, because hot fluids tend to expand. Being less dense, it is more buoyant than the overlying syrup, so it rises to the surface. The syrup at the surface, being cooler, has contracted slightly and is therefore denser than the underlying syrup. And so there is a tendency for it to sink. The contents of the saucepan will soon organize itself into cells with rising and sinking limbs—this is convection. The overall effect of all this is to transfer heat from the bottom to the top of the saucepan; in fact, this is a far more efficient way for the syrup to lose heat, compared to the way the interior of a rigid object like a hot brick can cool down.

At the end of the nineteenth century, the eminent physicist Lord Rayleigh investigated the conditions that are necessary for a fluid to undergo convection. He found that a number of factors are important. Of course, there is gravity. But how this acts depends on the temperature difference between the top and bottom of the fluid, the degree to which the fluid will expand as it gets hotter, and how sticky it is. In addition, there is also the thermal conductivity of the fluid, and the size of the tank or saucepan. Lord Rayleigh combined all these factors into a single number called the Rayleigh number. This number is really just a measure of the ratio between the gravitational force that drives convection and those forces that will resist it. The bigger the ratio, the greater the driving force and the less the resisting forces. Lord Rayleigh found that a fluid will always convect when the ratio exceeds about a thousand. In this case, it should be possible to test whether the Earth is convecting by calculating its Rayleigh number.

In the Earth, it is the mantle that acts like the hot syrup. It is heated by radioactivity and at its base by the molten outer core, which is at a temperature over 5000°C. And it is kept cool much nearer the surface by the atmosphere and oceans. Geophysicists have measured all the relevant properties of mantle rocks needed to calculate the Rayleigh number. The most difficult of these to estimate has been its stickiness or viscosity. To determine this, geophysicists have had to study the landscape in Scandinavia and northern Canada, regions that were covered by huge ice sheets about twenty thousand years ago. The enormous weight of the ice pushed down on the Earth's surface, creating wide depressions. But now, since the ice sheets have melted and dwindled away, as the climate has warmed, the floors of these depressions are slowly rising again; the speed at which this is happening is a measure of the viscosity of the mantle. A typical rate of uplift is about two centimeters a year, making the viscosity of the mantle about ten billion trillion poises (remember that the viscosity of water is about one-hundredth of a poise) — the mantle is clearly a pretty stiff fluid. Despite this, it is hot and large enough for its Rayleigh number to be a staggering ten million. In other words, the Rayleigh number of the mantle is abundantly supercritical, and we would expect — as, indeed, plate tectonics tells us — that it is convecting vigorously.

The reason the top of a convection cell in the Earth forms a tectonic plate — in other words, the lithosphere — is largely because, being cooler than the rest of the mantle, it is much stronger. For example, my own estimate of the viscosity of the lithosphere in South America beneath the Andes — relatively weak lithosphere that is now crumpling — shows that it is ten times more sticky than the underlying mantle. There are many lines of evidence that the lithospheres beneath the ancient cores of continents, such as the Brazilian Shield in South America, and much of the sea floor — true rigid plates — are even stiffer than this. This strength allows the vertical force of gravity, which drives convection, to be converted into horizontal forces that push and pull the plates. The push comes from the tendency — again, under the force of gravity — for the rising hot mantle beneath the midocean ridge, where the tectonic plates are created and move apart, to spread out sideways. The pull comes from the weight of the long slab of lithosphere, dangling deep in the Earth's interior at a subduction zone; it is transmitted along the

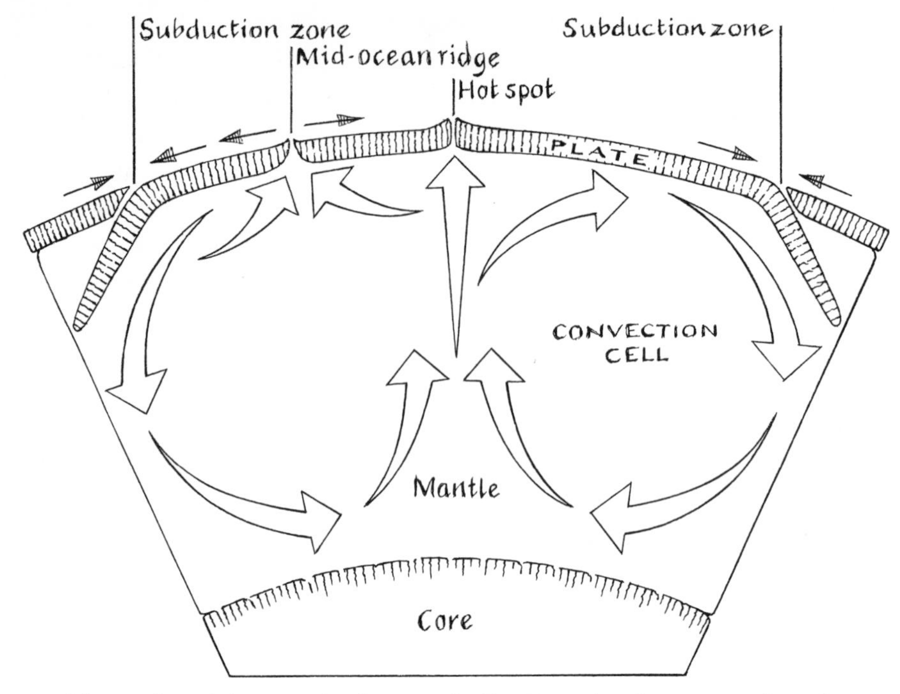

The motion of the tectonic plates on the Earth's surface is an expression of a much deeper flow, in which hotter mantle rocks rise up and colder ones sink in a pattern of convection. This convection is driven by gravity and also the heat from the core and mantle, and it is the way the interior of our planet loses heat and cools.

length of the slab, helping to move the plate horizontally, more or less in the same way that a toddler, clinging to an overhanging piece of a tablecloth, can drag the meal above across the table and down on top of him or her.

Getting Enough Push

The tectonic plates must be stronger than the push and pull that moves them. Otherwise, the plates would no longer remain rigid but would start to break or distort internally. However, as the rise of the great mountain ranges has graphically demonstrated, not all of the Earth's surface is covered by rigid tectonic plates. The rocks in the lithosphere of these nonplatelike portions of the world—places at the edges of plates, where the rocks really do break and distort—must always be weaker than those in the rigid plates themselves; this distortion and breakage is caused by the very same forces that drive the motion of the plates. The most obvious plate edges where this is happening are the gigantic inclined faults in subduction zones, such as the one along the length of the Andes in South America.

The force that has pushed up the Andes comes from the adjacent Nazca plate, applied through the action of this plate rubbing against the overlying rocks, as the Pacific ocean floor slips back into the mantle along the inclined fault. This rubbing generates a push on the Andes that depends on the slipperiness or lubrication of the fault. In much the same way, the maximum push you can apply to an object—for example, a broken-down car—depends on how well your feet grip the ground. If the subduction zone is slippery, then the maximum push across it will be small—this is like trying to push the car while standing on a patch of ice. On the other hand, if the subduction zone is very sticky or rough, it can transmit a much larger push. And it is this larger push that may be enough to squeeze and push up high mountains. It will, however, be opposed by two other forces. The first is the inherent strength of the lithosphere itself—its resistance to squashing or squeezing. And the second is the weight of the mountains—their tendency to fall down. These three forces are poised in a delicate balance.

The force balance essentially determines both the maximum height that a mountain range can reach and how rapidly it can be squeezed. This is easiest to see if we accept that the mountains really are capable of flowing like a fluid. Fluids respond to forces by flowing. The rate at which they flow for any given force is determined by their viscosity—hot runny syrup is easier to stir than cold sticky syrup. We can now begin to see the effect of different combinations of the forces that control the flow of the mountains. For example, if the subduction zone is rough, and the mountains have a low viscosity, then there will be enough transmitted push to squeeze them relatively rapidly; a slippery subduction zone will be

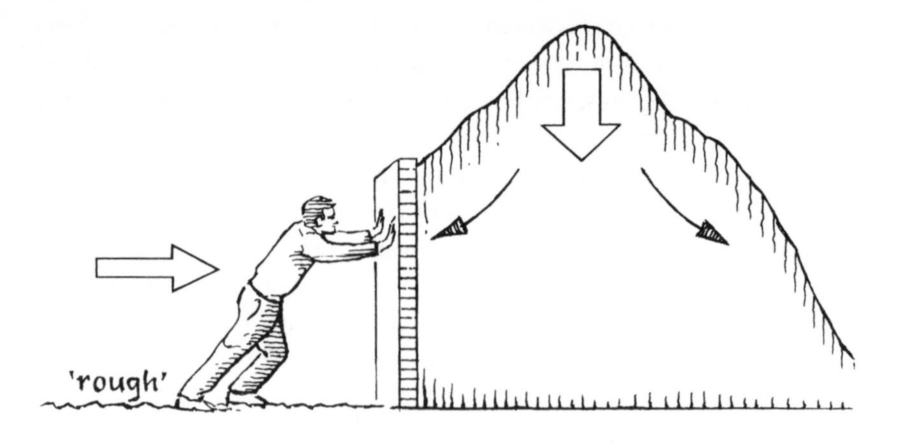

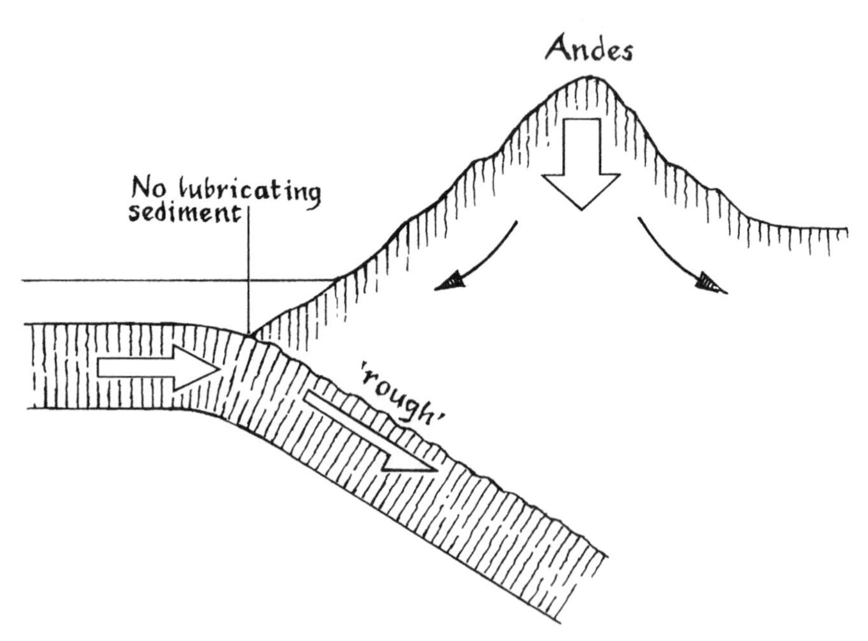

Mountains are held up by forces in the Earth, in much the same way that a mound of sand must be supported at its sides if it is not to collapse. In a subduction zone, the amount of push depends on the lubrication between the tectonic plates, as one plate slides beneath the other. If the plate interface is highly lubricated, perhaps by sediment, then the available push can hold up only relatively low mountains. A rough plate interface with no lubricating sediment could hold up much higher mountains—this seems to be the case along the subduction zone of central South America.

able to transmit sufficient push only to squeeze them more slowly. And the mountains will reach their maximum height when the push transmitted across the subduction zone balances their weight.

Putting these simple ideas together with all our geological discoveries in the Bolivian Andes, described in the previous chapters, we can attempt to answer the fundamental questions about the life cycle of these mountains, such as why they exist in the first place, why they look they way they do, and why they have evolved in their own peculiar way.

The Mountains Are Born

To a geologist, the most distinctive sign of subduction in the geological record is the presence of volcanic rocks with a similar composition—andesite—to those erupted today in the Andes. Ancient rock formations of andesite have been found along the western margin of South America; these are of all ages, dating right back to the remote Ordovician period, over four hundred million years ago. This is the best evidence that subduction has being going more or less continuously ever since. So, the first question is: Why did these mountains start to rise roughly forty million years ago? Or put another way, why did the push become sufficient to make mountains only at this time?

An answer may lie in the speed at which the two tectonic plates are approaching each other in a subduction zone. Geologists can track the plates through time, going back nearly two hundred million years, using the record of their creation and motion imprinted on the sea floor. This shows that around forty million years ago the Nazca and South American plates started to converge much more slowly, decelerating from nearly 150 millimeters per year to 50 millimeters per year. One reason this might have happened is that the Indian plate had started to collide with the great continent of Eurasia, slowing down dramatically its northward motion and forcing the direction and speed of many of the other tectonic plates to change as well. It seems that these effects were also felt in the long subduction margin along the western margin of South America, and here the two plates started rubbing against each other much less vigorously.

It may seem paradoxical that a slowdown in the plate activity could start the raising of the Bolivian Andes. But consider what happens when two sticks are rubbed together. The friction creates heat, sometimes even enough to start a fire. In the Earth, heat tends to soften or weaken rocks, allowing them to slip past each other more easily. In other words, hot subduction zones are not the places where we would expect much push to be generated. But if the motion between two plates slackens, there is less heat from their friction and the subduction zone starts to cool down and become stronger—such an increase in strength may have been enough to squeeze the continent to the east. However, I believe that this is no more than part of the explanation. There is something else, and it has nothing to do with the speed of the plates, but rather a change in the planet's climate. But first, let me ask some more questions.

Why are the mountains confined to the western margin of South America? Why did not the whole of South America buckle under the pressure of pushing up the mountains? The answer must be that the western margin of South America was, and is, inherently weaker than the rest of the continent. In terms of the fluid theory, this side of the continent has a lower viscosity. The Brazilian Shield, which occupies the heart of South America, forms a strong and ancient core to the continent which has been more or less unaffected by any significant geological event over the last six hundred million years. For this reason, it has had plenty of time to slowly lose heat, becoming steadily colder and stronger. In contrast, the western margin of the continent, close to the subduction zone, has been repeatedly warmed up by volcanic activity associated with the subduction zone and battered by a past history of squeezing and even stretching at times. It has therefore behaved like a much less viscous fluid than the Brazilian Shield. So when the push across the subduction zone dramatically increased around forty million years ago, squeezing was concentrated into the "soft" zone on the western margin of the continent.

We can look in more detail at the early response of the western margin of South America to the increased push. Our study of the river dunes preserved in the red-beds (see chapter 5) showed that at first two sibling mountain ranges started to grow. One followed the chain of volcanoes in the west. The other lay about three hundred kilometers farther east, rising up out of the deepest part—lowest point—of the old Cretaceous inland sea or lake. Our fluid theory again explains why this should have happened: the push that creates the mountains has to work against both the stiffness of the lithosphere and its tendency to fall down. The mountains will first start rising wherever these are least, because here there will be more push available for getting the lithosphere to flow. Thus, the western range began rising along a weaker zone in the thermally softened volcanic arc. The eastern range also nucleated along a weaker zone, but this was also a topographically low point where the tendency to fall down was least.

The Mountains Get Older

So what would we expect to happen to these two early fluid welts on the western margin of South America? If the push across the subduction zone is maintained, the mountains will continue flowing up. And our understanding of fluids shows us that they will do this by getting higher and wider. This is what the rock record tells us has happened in the Bolivian Andes. The ranges rose mainly as a consequence of the squeezing of the crust, although volcanic activity, both at the surface and much deeper down, has played a role, especially in building up the ranges in the west. As they got wider, the two sibling ranges reached out toward each other—they were much closer about twenty-five million years ago, while the eastern range continued to widen toward the heart of South America. And then, about ten million years ago, the fluidlike mountains began to impinge on the much stiffer Brazilian Shield. At this point, there was a dramatic change in the growth of the mountains—they surged over the Brazilian Shield along a gigantic, gently inclined fault, nearly doubling their height. The maximum height that these mountains are capable of reaching is determined by the balance between the horizontal push and the weight of the mountains. In fact, the high Andes are probably now close to their maximum height, at a general elevation of about 4 kilometers above sea level—an idea to which I will return.

The magnetism in the rocks (see chapter 7) shows that as the mountains rose, they also swung around, accentuating the curvature of the western margin of South America. This is really a consequence of more squeezing in the knee of the Andes—in Bolivia—than farther north or south, so that the coastline here migrated farther east, bending the leg of the mountains. But why should there be a variation in the amount of squeezing along the length of the Andes? The answer must be simply that the Andes are easier to squeeze—are softer or weaker—in some places than others. You may be confused now, because I have already invoked a softening or weakening of South America—a result of the earlier pummeling and heating it has received in its long geological history—to explain the birth and subsequent growth of the Andes on the western side of the continent. But one would hardly expect the effects of all this geological activity to be precisely the same everywhere, and as we can see in the bending of the mountains, the consequences of even quite small differences in stiffness along the mountains' length can be surprisingly profound. In some ways, we can turn that famous dictum "the present is the key to the past" on its head. When it comes to mountain building, it often seems that the past is the key to the present.

Mountains That Bob Up

So far, we have viewed the rise of mountains in terms of a fluid, squeezed and thickened. This squeezing is driven by a horizontal push. Yet there are some observations about both the Andes and the great ranges of Asia that suggest that something else is going on as well. For a start, there is a persistent discrepancy between the observed amount of squeezing in the Bolivian Andes and the depth of the roots in the crust; there does not seem to be enough squeezing to push down the roots that we can see. Our study of the volcanic activity in the high Andes has suggested a solution to this problem: a large part of the roots in the crust beneath the volcanoes is made of solidified magma that has risen up from the underlying mantle. However, the tell-tale signature of helium leaking out of the ground shows that the mantle is melting over a much wider region, extending for over three hundred kilometers east of these volcanoes. Such a wide region of mantle melting is a clue to another way of raising mountains, first discovered in the highest region of all, the great plateau of Tibet.

The Tibetan Plateau occupies a region of over two million square kilometers at a general elevation of about 5 kilometers above sea level. Despite its size, it has been largely unexplored by Western geologists. Political tensions have long made access difficult, and geologists have had to be content with studying the region from afar. They began by looking at images taken from space. These revealed what looked to be large rifts in the crust extending for hundreds of kilometers in a north-south direction. The plateau seems to be splitting apart. In the early 1980s, Western geologists were finally allowed to visit the region on the ground. They soon confirmed that the features in the satellite pictures are indeed rifts. They also found something else, possibly more surprising. There are signs of volcanoes throughout the plateau where a dark lava—basalt—has been erupted on the surface. In northern Tibet some of these volcanoes are active. Just as in the Bolivian Altiplano, all this suggests that the mantle beneath Tibet is melting.

In 1995 I had an opportunity to visit Tibet together with a team of geologists and geophysicists, including Philip England. I decided to collect and analyze the gases spewing out of the hot springs in the region, in much the same way that we had done in the Bolivian Andes. Our experiences in Tibet make a good story, but there is no room to tell it here. The main scientific outcome, as far as I was concerned, was the discovery that many of the hot springs were emitting helium straight from the mantle. We were able to pinpoint fairly precisely the region beneath Tibet where the mantle was melting, releasing its helium—it was melting beneath most of central and northern Tibet. More puzzling, the rocks showed that there has been very little squeezing of the crust in the last few tens of millions of years. So how did this region get so high?

The new discoveries in Tibet are a major geological problem. This region lies in the collision zone between the Indian and Eurasian plates, and it has long been assumed that the rise of these mountains is a direct consequence of the squeezing of the Tibetan crust, pushing the rocks together. In the early 1980s, Philip England had investigated the implications of this for fluidlike mountains. Though he could explain the overall extent of the Tibetan Plateau, he had considerable difficulty making sense of the great rifts in the plateau, where the crust was splitting or being pulled apart—quite the opposite to the expected pushing together. If we recall the basic force balance between the horizontal push and the weight of the mountains, it is clear that the only way that this can happen is if, somehow, the weight of the mountains becomes too great for the amount of horizontal push. The problem is understanding how the mountains could ever get into this situation. Fluid mountains would not be expected to rise beyond the point at which their weight equals the push. It was not until several years later that it dawned on England, together with his colleague Greg Houseman, that the answer might lie underneath the crust, in the mantle—a part of the lithosphere that is frustratingly hidden from geologists.

England and Houseman suggested that squeezing thickens not only the crust, but also the rest of the lithosphere in the mantle, creating a deep root. The root, being colder than the underlying portions of the mantle, is also heavier; given a chance it will sink into the Earth. But attached to the rest of the lithosphere, it is an anchor, holding the crust and mountains down. So, what happens if the mantle root falls off? The answer is that the mountains bob up without undergoing any further internal squeezing. Philip England calculated that they could easily bob up several kilometers, given the anticipated size of the mantle root. And there are other consequences to this bobbing up. To begin with, the mountains now become too high to be held up by the surrounding horizontal

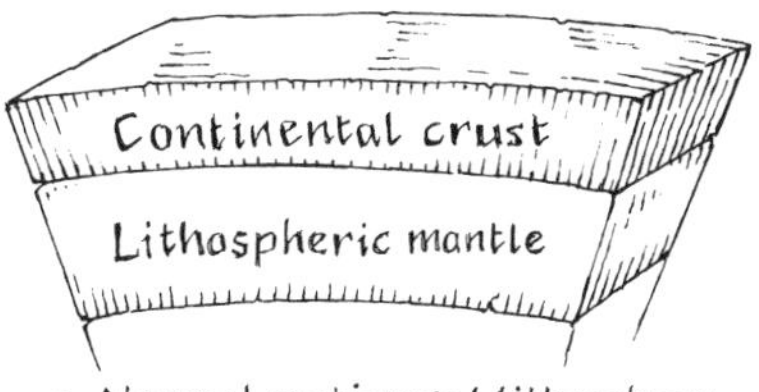

a. Normal continental lithosphere

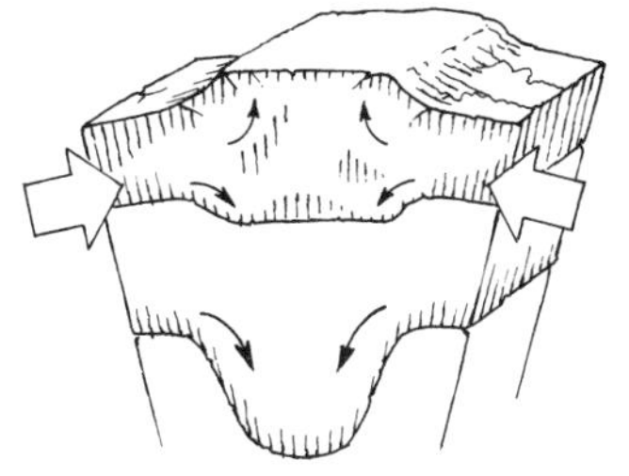

b. Mountains flow-up, root formed

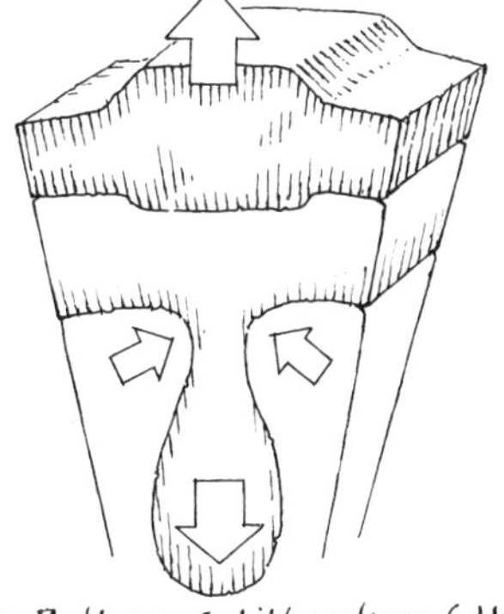

c. Bottom of lithosphere falls away
mountains bob up.

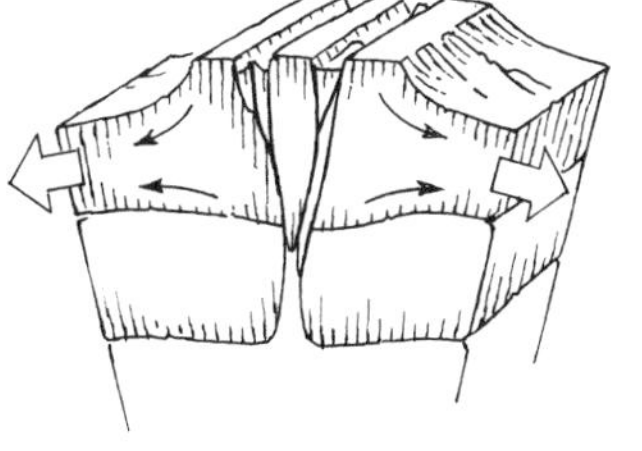

d. Crust extended, volcanic activity

Mountain building is more than just squeezing the Earth's crust—the underlying mantle in the lithosphere is thickened too, creating a cold, dense root that forms an anchor holding the mountains down. If the mantle root becomes too thick, it may fall off and sink deeper into the Earth's interior. This way, the anchor is removed and the mountains bob up. However, forces are set in motion that will lead ultimately to both the mountain's collapse and volcanic activity.

push. And so they start collapsing, spreading out sideways along rifts like the ones observed in Tibet. Beneath these rifts, in the mantle, much hotter rocks can then flow in to fill the void left behind by the relatively cold sinking mantle root. This compensating flow is hot enough to trigger melting in a wide region, creating a magma that will rise up into the crust, eventually erupting at the surface as basaltic lava flows and liberating its cargo of mantle helium.

Not all geologists are convinced. Though plausible, England and Houseman's ideas are also very speculative, relying on many unproved phenomena. Nonetheless, I have always been intrigued by the fact that there are many similarities between the Tibetan Plateau and the Altiplano in Bolivia. Fossil leaves in the Altiplano suggest that this region has risen in the last few million years without any significant internal squeezing. Indeed, there are signs, albeit very small, that this region is just beginning to rift apart. In 1945 an earthquake near Ancash in the Peruvian part of the high Andes was triggered by exactly this type of faulting in the crust. And there is abundant evidence from volcanism and helium that the mantle in a wide region beneath the Altiplano and part of the Eastern Cordillera is melting. So, could the Altiplano, like the Tibetan Plateau, also have lost its mantle root and bobbed up? It is not so easy to answer this question, but there is a suggestion that something like this, but not quite the same, did happen here.

Looking for Roots in the Mantle

The thickness of the crust beneath the high Bolivian Altiplano is about double the thickness of the crust beneath the Bolivian lowlands. Before the mountains started to rise the crust beneath the Altiplano must have been more or less the same as that beneath the lowlands. It is also likely that the thickness of the lithosphere was originally the same in both regions. Given that mountain building in the Andes is something that must affect the whole thickness of the lithosphere—both the crustal and mantle parts—a simple doubling of the crust beneath the Altiplano by squeezing should double the thickness of the lithosphere as well. Therefore, we would expect the lithosphere beneath the high Andes to be much thicker than that beneath the lowlands.

Seismologists have used earthquake vibrations to probe the thickness of the lithosphere beneath the Bolivian Andes. Earthquakes vibrations will be expected to travel faster through the more rigid and colder lithosphere than in the underlying parts of the mantle. So the base of the lithosphere should mark a change in the speed of these vibrations, though the change could be quite small and rather difficult to see. Undeterred, a group of American seismologists—

part of the same team that had already used earthquakes to determine the thickness of the crust beneath the Bolivian Andes—have scrutinized their seismic data searching for evidence of variations in the speed of the earthquake waves. They seem to be detecting a base to the lithosphere that is rather irregular, but nowhere is there any sign that it is significantly thicker beneath the Altiplano compared to below the eastern lowlands. In places, it is even much thinner.

Independently, we have found another way to estimate the thickness of the lithosphere. As part of a study on young volcanic rocks in the Altiplano, less than three million years old, we have made numerous analyses of the chemical composition of the lavas erupted from small volcanoes far to the east of the main chain of volcanoes. By chance, I found myself at a conference sitting next to Dan McKenzie from Cambridge University. He remarked that he was able to use the chemical composition of the lava to calculate the depth in the Earth from which they had originally come. The idea is that the composition depends on the type of minerals—garnet, for example—in the mantle rocks, deep in the Earth, when they melted to form the lava. And the presence of these particular minerals depends on the pressure that these rocks are under.

The pressure is simply a result of the weight of the overlying layers of rock—the deeper the rocks, the greater the weight and pressure. Thus, it is a relatively simple matter to calculate the pressure at any particular depth in the Earth. I gave Dan the chemical analyses of the lavas we had sampled. To my surprise, they contained a clear fingerprint of minerals that exist only at relatively low pressures, and hence shallow depths, in the Earth. In fact, Dan calculated that the original molten rock formed at a depth between 70 and 90 kilometers beneath the Altiplano. The temperature of this magma will be over 1000°C and close to that expected in the Earth at the base of the lithosphere. Despite the many uncertainties and assumptions, all this points to a lithosphere beneath the Altiplano that is around a hundred kilometers thick, a thickness that is surprisingly thin and not much more than the thickness of the crust itself in this region. East of the Andes, beneath the Brazilian Shield, the evidence is that the lithosphere is over two hundred kilometers thick. Somehow, like Tibet, the Altiplano has lost part of its mantle root.

A Twist in the Tale

I am not convinced that the Bolivian Altiplano has bobbed up in exactly the same way as that proposed for the Tibetan Plateau. There are many differences between the two regions. For a start, unlike the Tibetan Plateau, the Bolivian Andes lies above a subduction zone. It is also over a kilometer lower in average elevation. And we can explain the rise of the Eastern Cordillera during the last ten million years (see chapter 6) as a consequence of sliding up and over the edge of the strong core of South America, in the Brazilian Shield, along a gigantic gently inclined fault. This suggests that any evidence for rise of the regions even farther west, in the Altiplano, can be explained in terms of the same process. What seems to have happened is that the edge of the Brazilian Shield has acted as a sort of plunger, pushing and squeezing its way into hot and ductile rocks, deep in the crust and mantle beneath the Altiplano. This squeezing would effectively pump up the "lid" of the crust without signs of horizontal earth movements at the surface.

While this was going on, something else must have happened to the mantle root beneath the Altiplano. To understand its probable fate, one needs to step back and return to that classic image of the subduction zone in which the Nazca plate slides into the mantle at an angle of about thirty degrees beneath South America. As the sinking plate penetrates the Earth's interior, it drags down the surrounding rocks with it. Try pushing your finger into some sticky syrup. The syrup will be forced aside, around your finger, but because it is so sticky, some of it will also be pushed down with your finger. In the mantle, the effect of this is to create a circular eddy current above the sinking slab—a bit like a rotating wheel—that continually sucks in hot rocks from farther away. These hot rocks melt as they come into contact with water streaming off the sinking ocean floor, providing the long-lived supply of magma that feeds the overlying chain of volcanoes.

The eddy of hot rocks may have another effect. One idea is that it interferes with the squeezing and thickening of the overlying lithosphere, scraping or pulling off small portions of the mantle root. In this way, an uneasy equilibrium is established between the tendency for the root to grow as the lithosphere is squeezed and the tendency for it to be swept away by the underlying mantle eddy current. Small patches of lithosphere over a wide region can even melt as they come into contact with the hot eddy current, and it may be this melting that we are detecting in our helium studies. This opens up a new possibility for the behavior of the mountains. As I have described, there is a suggestion that Tibet, which lies in the collision zone between two continents, bobbed up several kilometers when it lost a large chunk of the root. In contrast, the high Bolivian Andes, which lie above the sinking plate in a subduction zone, seem to have been steadily rising over a long period of time as the crust was progressively thickened. But it would be hard to rule out a number of wobbles or jerks superimposed on this upward motion, each driven by the effect of the removal of a small portion of the mantle anchor. Perhaps the mountains have just experienced another small upward jerk in their overall rise?

Little Mountains and Big Mountains

Despite all the complexity of the details of the growth of the great ranges of Asia and the Andes, outlined above, there is still an overall simplicity in their fluidlike behavior. The continents can clearly be weak enough to flow up as mountains, a flow that is a response to the squeezing of the lithosphere by external forces, though this may set in motion other processes that will influence the mountains' rise. However, they clearly do not flow up to the same extent everywhere. For example, only in the Himalayas do peaks rise above 8,000 meters. The Andes have no peaks higher than 7,000 meters, but they contain many over 6,000 meters; mountains like these are rare throughout the rest of the world.

It is not just the heights of the highest points that determine the size of a mountain range, it is also their sheer bulk. One can see this more clearly if one imagines smoothing out the high peaks by bulldozing them into the adjacent deep valleys, creating a general bulge in the Earth's surface. One can then get a measure of the size of the bulge of the Andes and the ranges in Asia by considering the area of land that lies higher than some well-known peaks, such as the Eiger—one of the higher peaks in the European Alps—or Mount Shasta in California. Well over two million square kilometers of the combined ranges of Tibet and the Himalayas, and over half a million square kilometers of the Andes, are higher than this, over 4,000 meters above sea level. Clearly, there must be sufficient forces in the Earth where the tectonic plates converge to push up such large mountains. So, first, why is it that just these two regions are so high and vast—no other mountain range on Earth comes close—given that there are many places where two plates are converging? And, second, why are the great ranges of Asia, in the Himalayas and Tibet, so much larger than even the Andes?

The second question is the easiest to answer. It is simply because the Andes are above a subduction zone, whereas the Himalayas and Tibetan Plateau lie between two colliding continents. Continental collision cannot be absorbed by subduction, with one plate just sliding beneath the other into the mantle. This is because the continents contain a large thickness of low-density crust and are not heavy enough to sink back into the Earth's interior—in subduction zones, it is always the much denser oceanic plate that is subducted. Instead, the two continents will crumple up, pushing up vast mountain ranges if the convergence continues for long enough—and, as we have seen, they may bob up even higher if the giant anchor of the mantle root, created by this immense amount of squeezing, drops off. For example, the collision between India and Eurasian started about fifty million years ago, not long before the Bolivian Andes began to rise. Since then, India has drifted a further two thousand kilometers northward into Eurasian. This northward advance has been almost entirely absorbed by squeezing and thickening of the continents. However, during the same time, the Nazca plate has glided nearer twenty-five hundred kilometers toward South America. Only a small fraction of this motion — less than one-fifth, amounting to less than five hundred kilometers — has been absorbed by squeezing in the Andes. Most of the motion has resulted in the Nazca plate merely slipping into the mantle beneath the Andes.

The great ranges of Asia are part of a more extensive belt of mountains — the Alpine-Himalayan belt — extending east through the Middle East and Turkey to the European Alps and Pyrenees beyond. These mountains follow the boundary between the tectonic plates of India, Arabia, and Africa in the south, and Eurasia in the north: a gigantic battle front between colliding continents. The front, however, is not all continental collision. It is broken in two by a length of subduction zone following the northern edge of the Indian ocean, along the coasts of Pakistan and Iran. There is another subduction zone to the east of India, extending for thousands of kilometers into Southeast Asia. Indeed, it is likely that the pull of the heavy, sinking slabs in these two subduction zones provides the main driving force for the collision of the plates. However, rather like massed armies in the field of battle, this force is more concentrated along the shorter front in northern India, but more spread out and weaker along the longer African and Arabian fronts, in the Mediterranean and Middle East. This is why the ranges of the Himalayas and Tibet are the highest of all; here more of the force needed to push up mountains is available. Away from the Alpine-Himalayan belt, the only other major site of continental collision is in the Southern Alps of New Zealand. But these mountains are too youthful to have had time to grow to any great size.

We can now return to our first question, recasting it in a slightly different way: Why are the Andes so much bigger than any other mountain range lying above a subduction zone? The answer goes back to the push that can be transmitted across the subduction zone. This will be strongly controlled, as I described earlier, by the slipperiness of the inclined fault along which the sinking plate slips into the underlying mantle. This slipperiness may be governed by a surprising factor—the availability of some sort of lubricant. The most readily available lubricant is the water-rich sediment that covers and smoothes out the deep ocean floor. Some of this sediment and water will be dragged down with the sinking plate and smeared out along the inclined fault, making it quite soft and slippery. Where the ocean floor is rugged and bare volcanic rock, the inclined plane is likely to be hard and rough. If we look at the ocean floor sinking beneath South America, we find that, indeed, much of it is like this; the deep ocean trench is virtually empty of any sediment. This observation may go a long way to explaining why there was a sufficient push to create the Andes. In other words, the presence of sediment in a trench can have a large effect on the size of the mountains. We can see this clearly when we look at the trenches of other subduction zones, such as along the North Island of New Zealand or the west coast of North America as far as southern Alaska. Here, there are huge piles of sediment, which, I believe, are lubricating the sinking ocean floor, making its passage more slippery as it slides into the Earth's mantle, thereby greatly reducing the forces that can be locally harnessed to build a mountain range.

It is not enough, though, for the subduction zone along the western margin of South America to be rough everywhere, because then the forces available to push up the Andes would be effectively diluted, spread out along a front over five thousand kilometers long—something else in the system would give before the Andes reached their present great height. However, restricting the rough patch to a relatively short section of the subduction zone, perhaps only a few thousand kilometers long, would focus the forces at work, concentrating them to such an extent that they could push up a range as high as the Andes in Bolivia. This is a bit like the difference between trying to push a sharp or blunt stake into the ground: the sharp stake goes in easily because your push is focused down on a narrow patch of ground beneath the stake's sharp point; the blunt stake is more difficult because the same push is now spread over the wider blunt end. A focusing of the push seems to be exactly what has happened in the central part of the Andes (in southern Peru,

Bolivia, northern Chile, and northern Argentina), because it is only here that the trench is empty, starved of its supply of lubricating sediment. This comprises about half the total length of trench along the western margin of South America—the rest, farther north and south, is swamped with lubricating sediment. So, perhaps an even more fundamental cause of the great height of the Bolivian Andes is the supply and distribution of sediment. In this case, we should ask what controls this pattern of sedimentation. And the answer to this turns out to be the climate.

The reason there is very little sediment in the trench along the central part of the Andes is that the coast here, in the Atacama Desert, is one of the driest places on Earth. There is virtually no rain to create rivers that could erode rock detritus from the mountains and carry them into the trench. The region is so dry because of a peculiar combination of circumstances. The unusually cold sea currents along this coast—the Humboldt or Peru-Chile Current system—are particularly important, because cold water does not evaporate so readily, and the overlying chilled air is too cold to rise—only rising masses of moist air form rain. In this way, the supply of moisture needed to create rain clouds in the Atacama Desert is cut off. Here, we have a direct link with the theme of climate that we started out with at the beginning of the chapter, provoking the extraordinary thought that the temperature of the water in the oceans can ultimately control the raising of large portions of the Earth's surface.

The link with climate is even more profound. First, it makes the rise of the mountains peculiarly sensitive to the atmospheric conditions at the Earth's surface. Changes in this, for whatever reason, could bring more of the available forces in the Earth to bear on a relatively short stretch of the margin between two colliding plates, causing a sudden boost in the elevation of the adjacent mountain range. But, second, as the mountains rise, they, in turn, influence the weather themselves, moderating the distribution of rainfall on their flanks. The eastern side of the Andes, in Peru and Bolivia, is unusually wet because the mountains deflect the westward-flowing air upward, causing this air to cool and release its moisture on the

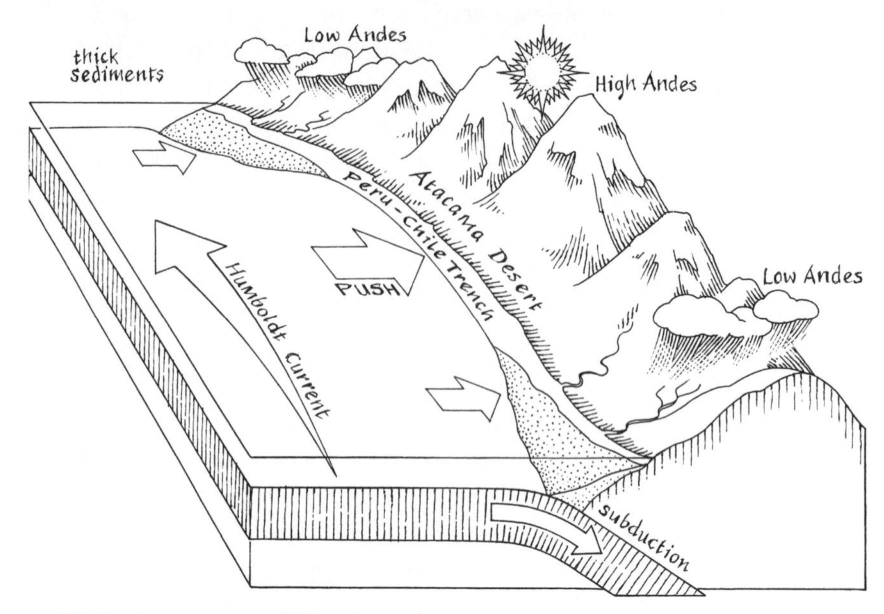

The high mountains of the Bolivian Andes may owe their existence to a peculiar feature of the climate. Cold currents—such as the Humboldt Current—sweep along the west side of South America, chilling the air and helping to make the coasts of Peru and Chile a desert. Here, virtually no sediment reaches the deep ocean trench offshore, and the sinking plate grinds against South America, providing enough push to hold the mountains up high. Farther south and north, where the climate is wetter, there is enough lubricating sediment to make the subduction zone too slippery to push up high mountains.

foothills of the Sub-Andes. And, as we have seen, the western flanks are extremely dry because the cold ocean prevents any rain coming from the west, and the winds from the east have lost most of their moisture in the Sub-Andes. All this means that the rivers flowing off the wet eastern flanks have far more energy, cutting down and wearing away the bedrock almost as fast as the rocks are pushed up. In other words, the mountains on this side of the Andes—a rugged landscape of high peaks and deep valleys—have been heavily sculpted by the forces of erosion. But the dry western margin of the Andes is smoother and more closely controlled by forces—the rise of magma and squeezing of the crust—originating deep inside the Earth.

A Cooling Story

Climatologists now believe that the rise of mountain ranges in the Andes and Himalayas has influenced the whole planet's climate over the past few tens of millions of years. And this is in the very period when primate evolution forged ahead from nocturnal tree-dwelling mammals to homo sapiens. The evidence lies in the composition of the calcareous shells of microscopic organisms called foraminifera.

Foraminifera live at various depths in the sea, and they make their shells by extracting the necessary chemical components from the water. The shells are made of calcium carbonate—limestone—containing calcium, carbon, and oxygen. Now there are several isotopes of oxygen, including light oxygen, oxygen-16, and heavy oxygen, oxygen-18. It turns out that when the foraminifera take oxygen from the water to build their shells, the amount of heavy oxygen that ends up in the shell depends on two factors. The first, most obviously, is the heavy oxygen concentration in the sea water itself, which, in turn, depends on how much sea water has evaporated to be locked up in ice sheets on land: the more ice, the more heavy oxygen in the ocean. But the amount of this heavy oxygen that actually gets into the shell is also controlled by the temperature of the sea water: the colder it is, the greater the amount of heavy oxygen.

This double significance of the oxygen isotopes means that when the organisms eventually die and accumulate on the sea floor, they leave a record in their shells of both ocean temperature and ice sheets through time. Geochemists can now read this record by drilling into the sea floor and collecting samples of the shells: these are then analyzed, using a mass spectrometer, to determine the proportion of heavy and light oxygen. In this way, it has become clear that over the past fifty million years or so, ice sheets—such as those in Antarctica—have expanded enormously, and the temperature at the bottom of the deep ocean has dropped over 15°C. These changes reflect a marked cooling of the planet's climate, a cooling that can also be seen in fossil plants and other organisms.

The remarkable fact is that the start of the general cooling of the global climate coincides with the onset of the rise of the Himalayas and Tibet (and the Andes too). Can there be a connection? In 1992 two American climatologists, Maureen Raymo and William Ruddiman, gave substance to a much older idea by suggesting that there is, and they proposed a mechanism by which mountains can cool the climate. The key to their idea is the fact that carbon dioxide is a greenhouse gas. This means that the presence of carbon dioxide in the atmosphere makes the atmosphere warmer than it would otherwise be, because carbon dioxide strongly absorbs the wavelengths of solar radiation that most effectively heat the atmosphere. It acts along with other greenhouse gases such as water and methane as a sort of thermal blanket draped around the planet, keeping out the icy coldness of space.

It follows that if the concentration of carbon dioxide in the atmosphere increases, the global climate will become, on average, warmer. And if it drops, the global climate will become colder. Climatologists believe that fluctuations in the concentration of carbon dioxide and methane in the atmosphere have been responsible for the major climatic events in the planet's history. When the level of these gases was low, the planet was exceptionally cold, gripped in the icy hold of an ice age, when ice sheets covered much of the continents. But when the level of greenhouse gases was high, the planet has been unusually warm. The last time that this happened was during the long summer of the Jurassic and Cretaceous periods of geological time, when the cold-blooded dinosaurs thrived. Since the end of the Cretaceous, the climate has steadily cooled, so that in the last forty million years or so the planet has been in an ice age. The logical conclusion is that this is because the level of greenhouse gases such as carbon dioxide have declined.

Raymo and Ruddiman suggested that the reason the level of carbon dioxide in the atmosphere has dropped is that it has been sucked out by chemical reactions between air and rocks in the mountains. The reactions are well known. There is always a certain amount of air, including carbon dioxide, dissolved in rain water. When the rain cascades down the sides of high mountains, it has enough energy to carry rock fragments with it. These rock fragments contain minerals called silicates. Silicates are minerals that are composed mainly of silicon and oxygen, and they crystallize deep inside the Earth at high temperatures and pressures. At the Earth's surface, where the temperatures and pressures are much lower, they are chemically unstable. It is therefore not surprising that rock silicates will react with water and carbon dioxide in the atmosphere to form a very common type of mineral called carbonate. When the rivers wash the carbonate down to the sea, it is quickly taken up by growing marine organisms to build their shells. Eventually, when these organisms die, they are buried on the sea floor and ultimately become limestone. The overall effect of this entire process is to take carbon dioxide out of the atmosphere and lock it up in rocks.

If there were no mountain building, then all the rocks at the Earth's surface would react relatively quickly with the carbon dioxide to form carbonate. At this stage the reactions would stop because there would be no more silicate left at the surface to fuel the chemical reactions. In this case, some sort of equilibrium would be reached, and the level of carbon dioxide in the atmosphere would remain more or less constant. Mountain building changes all of that. As long as highlands exist, rivers will flow off them, wearing away the mountains. In this way, new rock with its silicates is constantly being exposed at the surface, ready to react with the atmosphere and remove that all-important carbon dioxide. Continued mountain building keeps this process going, acting as a sort of rock pump that first sucks carbon dioxide out of the atmosphere and then locks it up in the oceans as limestone. And the rock pump will be most active when large mountain ranges are growing, providing a continuous supply of new rock that can be worn away by rivers. In fact, the pump may be too active unless there is some other process, as yet poorly understood, that undermines its efficiency by returning some of the carbon dioxide back to the atmosphere—it has been calculated that the pump unhindered could suck out all of the atmosphere's supply of carbon dioxide in much less than a million years.

The mountains of Central Asia have added enormously to the capacity of this pump. Just how much is clear when one considers the amount of rock that has been removed by erosion from these mountains to be dumped on the sea floor. The rock washed off the high Himalayas is mainly carried by the Ganges down to the delta region of Bangladesh. Here it is added to the enormous pile of sediment, several kilometers deep, which extends underwater hundreds of kilometers out to sea, in the Bay of Bengal. But the rivers of the Indian subcontinent are not alone in this sediment transport. Rock is also carried from the Andes down into the Amazon River and eventually is deposited in the huge Amazon underwater fan off the coast of Brazil. In fact, three quarters of the material carried by the Amazon at its mouth, after a four-thousand-kilometer journey, comes from the eastern slopes of the Andes of Bolivia, Peru, Ecuador, and Colombia.

Raymo and Ruddiman's theory suggests that the rise of the Himalayas and Tibet has been directly responsible for the steady cooling of the planet's climate over the last few tens of millions of years. And, I believe that this has had many far-reaching effects—I finally return to my other explanation for the trigger that set off the rise of the Andes in the first place and guided their subsequent growth. The most dramatic shift to cold conditions occurred in the polar regions. Today, the oceans here are always close to freezing—each winter, it is cold enough for even the salty sea water to turn to ice—and tongues of this very cold water find their way into the rest of the world's oceans. So it is the water from around Antarctica that is ultimately responsible for making the Pacific off the west coast of South America so unusually cold.

I have already described how this unusually cold water has helped to turn the coast of Chile and Peru into the driest place on Earth. But this was not always the case: when the oceans were much warmer, the coast of Peru and Chile was very likely a lot less arid. So when did this all change? It must have been when the planet started its long slide into an ice age—around forty million years ago—with the growth of the vast Antarctic ice sheets and a deepening chill in the oceans. It seems too much of a coincidence that this is the very moment when the Andean mountains in Bolivia first started to rise. I believe that the explanation is quite simple but extraordinary nonetheless. The cooling of the oceans caused the rivers of Peru and Chile flowing into the Pacific to dry up, turning off the supply of lubricating sediment to the trench and subduction zone offshore, and turning on the push needed to raise the mountains. And if this cooling was ultimately set in motion by the rise of the Himalayas and Tibet, then we have found the fascinating link between the two great mountain ranges on the planet. In step, like a sort of intricate tango (the Latin connection, perhaps?), they have danced together.

Elsewhere in the world, the cooling of the climate was a major force in the long-term evolution of grasslands. These grasslands were exploited by grazing mammals as an abundant source of food. In this way animals and plants were locked into a course of evolution that has given us much of the fauna and flora of the planet today. During the last ten million years or so, there has been a shift toward a climate that is much more seasonal, with greater contrasts between warm and cold, and wet and dry, conditions. Anthropologists have long thought that the progressive spread of grass savannah at the expense of tropical jungle, driven by this increased seasonality in Africa, may have played a pivotal role in the evolution of primates. The shrinking of the rain forests may have provided that window of opportunity for these primates, with their delicate grasping hands and forward-looking eyes, to leave the trees and live on the plains. This change ultimately led to the start of own evolution, between five and ten million years ago, when hominids branched from the great apes. And looking much further back in time, the rise and subsequent demise of long-forgotten mountain ranges may have had their own profound influence on both the planet's climate and the earlier course of evolution.

There is one final, important question about the life of mountains. Why and how do mountains die?

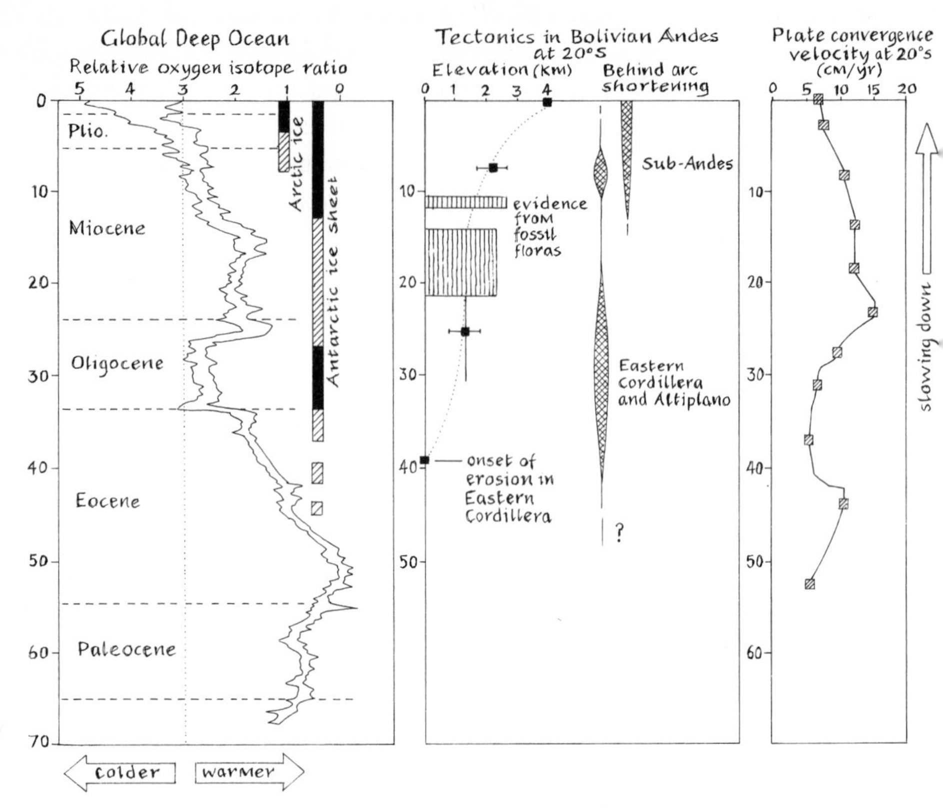

Many factors may have been responsible for the rise of the high Andes during the last few tens of millions of years. Phases of mountain building coincide with a slowdown in the convergence between the Nazca and South America plates. But the progressive cooling of the world's oceans and the growth of ice sheets—recorded in the oxygen isotopic composition of fossil shells—seem to show a more striking link with the history of squeezing and uplift of the Bolivian Andes. This cooling has helped to make the west coast of Peru and Chile unusually dry. With no rivers to carry rock debris to the Pacific, the trench offshore is deprived of lubricating sediment. This way, the subduction zone stays rough enough to push up high mountains.

Death of a Mountain

If you look at a geological map of virtually any part of Britain, you will find that its colorful patterns contain the evidence for a history of upheavals in the Earth's crust, "revolutions still more remote in the distance of this extraordinary perspective," as John Playfair expressed it at the end of the eighteenth century, after viewing Hutton's Unconformity in Scotland. These revolutions seem to be an alternation of uplift and subsidence of the planet's surface—the life and death of mountain ranges. In fact, if we equate every million years of a mountain range's life with a year of our own, then we would not be far wrong in saying that mountains do not live much more than three score and ten.

We have seen how mountains like the Andes have been pushed up, eventually reaching a maximum height when this push balances the weight of the mountains. The long-term survival of these ranges depends on the ability of the forces that pushed them up to counteract the effects of erosion. This is clear when we consider that most mountain ranges have had many times their present height stripped away by the relentless activity of rivers, as they carry rock fragments to the lowlands. These rivers bite deep into the bedrock, creating wounds that can be healed only by the continued rise of the mountains themselves. Remove the push and two things happen: the mountains start to collapse under their own weight, and the rivers wear away what is left. In time, the once high and mighty mountain range is reduced to a landscape of low and rolling hills.

The push that holds up the mountains may decline for a number of reasons. Perhaps a global reorganization of the tectonic plates causes their relative motions, at any particular plate boundary, to change; eventually the push may even turn into a pull and the mountains will rift apart, creating a new ocean. Perhaps, in the same way that the past cooling of the climate may have turned off the supply of lubricating sediment in the trench along the coast of Peru and Chile, so a future warming—who knows when—will turn it on again? In this way, the subduction zone will become too slippery and unable to hold up the high Andes—this great pile of rock will start collapsing, falling apart as it sinks to a much lower level. The new ideas about the rise of Tibet suggest another, even more powerful, way that a high mountain range might be brought down. Tibet may have bobbed up when the anchor of the mantle root fell off. But by becoming too high, the mountains trigger the destructive force of gravity—manifest in the present rifting or splitting of the crust—that is not only now reducing the height of the Tibetan Plateau but, it turns out, could eventually bring it down below sea level.

So perhaps Tibet and the high Andes, now several thousand meters above sea level, could be underwater a few tens of millions of years from now? Ancient unconformities in the rocks suggest that it has happened before. Perhaps it has already started? One can certainly single out other mountain ranges that we know are dying—parts of the European Alps and North American Rockies are old men (or women), now collapsing or being rapidly worn away, and the remnants of the Appalachians or the Urals belong to many generations past. As these high features of the planet gradually disappear, they will be superseded elsewhere by the youthful, sinuous mountain ranges of Eurasia, stretching from northern Turkey to the Himalayas, as well as many of the small mountain ranges around the rim of the Pacific Ocean, including the Southern Alps in New Zealand.

The rise and fall of vast mountain ranges are really expressions of the fundamental fluid nature of our planet. Deep in its interior, the churnings of the hot mantle rocks, as they flow in a pattern of convection, drive the motion of the plates and provide the force to push up mountains. And the flow of the mountains themselves is part of the fluidlike behavior of the Earth's surface. But the force of gravity, whether through the collapse and lateral spreading of the fluid mountains, or through the flow of water downhill, will ultimately destroy them. All this reveals mountains as dynamic features of our planet, deeply intertwined, as we have seen, with the Earth's climate and, ultimately, living organisms themselves.

Making Ends Meet

I have now finally reached my fundamental conclusion that mountains, with their life and death cycles, are almost living creatures, and, much like members of a close-knit family, their life stories are deeply felt by the rest of the planet. I hope I have shown in the previous chapters of this book how this lifelike behavior of the Andes can be seen in the remains of ancient rivers, forever seeking out new paths to the lowlands, or in the contorted and faulted layers of rock, squeezed and pushed by the relentless movements in the crust, or in the constant volcanic activity that has covered the landscape with ash and built up huge bodies of molten rock at depth. I now draw close to the conclusion of my own personal search for the origin of these mountains, a search that has taken me far beyond South America. Sometimes, I have felt like a cave explorer, casting a beam around in an underground cavern, pinpointing in the shaft of light marvels of this strange world, hoping to startle the Bolivian miners' devil that lurks there, and trying to illuminate the deepest levels in our understanding of mountains. There are still many dark areas that I have failed to light up. But I hope that I have made clear at least one geologist's view of these extraordinary parts of the Earth, opening up a new way of thinking about the surface of the planet.

The research has taken me over ten years, and during that time I have spent almost three years in Bolivia, visiting virtually all of the high mountainous parts of the country. If the tachometer of our Land Cruiser is to be believed, we have traveled well over 150,000 kilometers on mainly rough dirt roads, camping out during the cold Andean nights. I have climbed many of the high active volcanoes and hacked my way through the thick undergrowth of the subtropical jungle in the foothills. I would like to end my narrative with some thoughts on whether all this effort was, in the final analysis, worth it. A litmus test of this, perhaps, is whether I can interest other people—not just geologists—in what I have done. This has been one of the main motivations behind writing this book. If one cannot make the geologist's view of mountains—

or, for that matter, any scientist's work—engage a wide audience, then, in my view, it is not just a failure of writing technique, but a challenge to the whole purpose of academic research.

I have had moments of more practical justification. Not long ago, I was contacted out of the blue by a geologist from an American oil giant. They had the license to look for oil in the high Altiplano—a license our original project sponsors once owned but had relinquished in the early 1990s when the price of crude oil plummeted. I was pleased to hear that the Bolivian Altiplano had not been abandoned altogether by the oil industry. I felt that I had come a long way since those early days at the start of our project, and I now knew enough about the geology of this region to be of some help. The company wanted an independent geologist to review their work and tell them if it made sense; they invited me to visit their headquarters in Texas, to speak to their geologists. And so it was that I found myself flying first class to the United States, armed with a thick stack of diagrams for my presentation. On arrival, I was overwhelmed by the heat and humidity of the place; city life consisted of moving as quickly as possible from one refrigerated box to another. Inside an air-conditioned steel and glass tower, I was introduced by one of the vice-presidents—there are lots of them—to the Bolivian project team as "Professor Simon Lamb from Oxford University, who will be with us for a few days—please take the opportunity of speaking to him." I was left to sift through geological reports and seismic data, before having intense discussions with a whole succession of geologists, geophysicists, and senior managers. I felt that I was at the nerve center of a powerful industrial machine, an unusual sensation for an academic geologist.

I recently had another experience that made me think more about the wider purpose or use of our Andean work. I was called by a television producer who was making a science program for BBC television. It was about a recent claim that the lost city of Atlantis had been finally found. To my amazement, the producer went on to say that it was now thought that Atlantis was in the high Bolivian Altiplano, nearly 4,000 meters above sea level. It turned out that a well-known Atlantis speculator—I forget his name—

had been looking at images of the Altiplano taken from space. He had seen long, straight lines filled with water that appeared to extend from the existing lakes in the region. It seemed to him that there could be only one explanation for these features: they were gigantic, man-made canals built by an ancient civilization that lived around the lakes. This civilization must have had enormous skills in taming large bodies of water, precisely fitting Plato's description of the works of the superhuman Atlanteans. The television producer had somehow heard about my geological research in Bolivia and wanted to know what I thought.

It took me a while to order my thoughts coherently. I certainly was not aware of any large, man-made canals in the Altiplano, a region that, by now, I knew extremely well. But I had satellite images of my own. So I offered to have a look and report back. The moment I put the receiver down I rushed to my stack of images. Very quickly I found the "canals" and laughed out loud. The reason I have never noticed them was because, as a geologist, I had instinctively interpreted them for what they are. The layers of strata in the Altiplano have been tilted slightly by movements of the crust, so that they form inclined slabs of rock dipping into the ground and intersecting the land surface as a series of parallel lines. Some layers are more resistant than others; the least resistant layers have been preferentially weathered out, leaving behind straight, parallel grooves. During the wet season, these grooves often fill up with water, looking like a network of straight and parallel man-made canals. There was absolutely no doubt in my mind that this is what we were looking at—a simple geological phenomenon that left no room for superhuman Atlanteans.

I contacted the TV producer and gave him my explanation. As I spoke, I sensed a lack of warmth in his response—he was distancing himself from me by qualifying my remarks with comments like "Well, that's your explanation, anyway" or "They could still be canals, though." It dawned on me that he wanted them to be Atlantean canals! So I ask myself: was it a failure or success to be able to demolish one person's Atlantis theory? Which is more interesting: bedrock reality or airy sensation? I would say that there is more at stake here than sensation; if we want to be able to distinguish between what is real and imaginary, useful and useless, so that we can make sense of our environment and have a better chance of surviving on this planet, then we have to be prepared to observe closely all aspects of the natural world, and that includes the Andes of South America.

Perhaps, though, it was the close collaboration and communication with other human beings that was the really important aspect of our scientific endeavor. The project brought together for a few years a small band of geologists who are now dispersed throughout the world, armed, though, with the experience of research in the Andes. Lorcan completed his thesis and gained his doctorate at Oxford, going on to work in the oil industry. Leonore took up a research post in New Zealand. And there were many other scientists with whom I have been lucky to work. However, we are all still in contact, still writing up the research for publication in scientific journals. I believe that we have also begun to understand better the Bolivian people themselves, in their long and hard struggle to survive in this hostile environment.

✣ Selected Glossary ✣

Acidic rock—Igneous rock that contains more than 63 percent by weight silica—granite or rhyolite are typical examples

Altiplano—The broad central part of the Andes in Peru, Chile, Bolivia, and Argentina, forming a high plateau with subdued relief at a general elevation of about 4,000 meters above sea level

Andesite—A fine-grained volcanic rock, named after the volcanoes of the Andes, composed mainly of feldspar, pyroxene, and sometimes quartz, amphibole, and olivine—its composition is defined as intermediate

Anomaly—Departure of some measured physical property (strength of magnetic field, acceleration due to gravity) from an expected value

Asthenosphere—The weak part of the mantle that immediately underlies the lithosphere

Basalt—A fine-grained volcanic rock composed mainly of feldspar, pyroxene, and sometimes olivine—its composition is defined as basic

Basic rock—Igneous rock containing 45 to 52 percent by weight silica—basalt is a typical example

Bedding—Layering in sedimentary rocks, caused by variations in the nature of the sediment, defining the approximate original horizontal

Benioff zone—An inclined zone of earthquakes in subduction zones

Black smoker—A column of hot, muddy water that gushes out of the sea floor

Cenozoic—"time of young life," which spans the present to sixty-five million years ago

Continental crust—Outer layer of the solid Earth beneath the continents, which is on average about thirty-five kilometers thick (but thicker beneath mountainous regions) and is rich in the minerals quartz and feldspar

Convection—Pattern of flow in a cooling fluid in which hot parts rise and cold parts sink

Core—The central part of the Earth where the Earth's magnetic field is generated, composed mainly of iron and nickel; the inner core is solid, but the outer core is molten

Crust—Outermost layer of the solid Earth, with a different composition from that of the underlying layers

Curie temperature—Temperature above which a magnet loses its permanent magnetism

Dune—Mound of sand with regular shape (steep lee face) created by a flow of water or wind

Earthquake—Vibration of the Earth caused by sudden slip (elastic rebound) on a fault

Eastern Cordillera—A wide, rugged mountainous region in Peru and Bolivia that lies east of the Altiplano—individual peaks reach 6,500 meters above sea level

Elastic rebound—Property of springiness of the Earth's crust, so that during an earthquake the crust "snaps" back like a stretched rubberband

Fault—Break in the Earth's crust—sudden movement on a fault triggers an earthquake

Feldspar—A silicate mineral that is common in igneous rocks

Fission track—Scar or track in crystal created by the spontaneous fission of uranium

Fluid—A substance that is capable of flowing—solid materials can be fluids

Free air anomaly—The difference between the measured value of the acceleration due to gravity and a standard reference value, taking account of the gravitational effects of latitude and elevation

Fumarole—Gas vent on a volcano, usually emitting superheated steam and other vapors

Granite—A coarse-grained igneous rock, commonly found in the continental crust and composed mainly of the minerals quartz, feldspar, and mica—its composition is defined as acidic

Gravity—The effect of the mutual attraction of matter, causing bodies to accelerate toward each other, hence gravitational acceleration (g)

Igneous rock—Rock that has cooled from a molten state

Ignimbrite—Compacted volcanic deposit that was originally blasted out of a volcano as a ground-hugging cloud of superheated ash and gas, traveling at high speed for large distances before settling out

Intermediate rock—Igneous rock with a moderate content of silica (52–63 percent by weight)—andesite is a typical example

Isotope—One of two or more forms of an element differing in atomic weight

Isostasy—The principle of a crust floating on a fluidlike mantle according to Archimedes principle

Lava—Molten rock (or solidified remains) that flows out from a volcano during a volcanic eruption

Lithosphere—Strong outer part of the solid Earth that forms a plate—contains both crust and part of the underlying mantle and is usually about one hundred kilometers thick

Magma—Body of molten rock

Mantle—Portion of the Earth between the crust and core—the top of the mantle is composed mainly of the minerals olivine and pyroxene

Mesozoic—"time of middle life"—period in Earth history between 65 and 250 million years ago

Metamorphic rock—A rock that has undergone changes due to the effects of temperature and pressure after the rock first formed

Midocean ridge—A linear zone of shallowing in the middle of oceans—new ocean crust is created by volcanic eruptions along the crest of the midocean ridge

Moho—Name for the base of the crust, marking the boundary with the underlying mantle

Oceanic crust—Crust beneath the oceans that is about seven kilometers thick, consisting of three principal layers: a top-covering thin layer of sedimentary rocks, a middle layer of fine-grained volcanic rock (basalt), and a bottom layer of coarse-grained igneous rock

Ocean trench—Long, deep depression in the ocean floor where a plate bends down and sinks back into the Earth's interior

Olivine—A silicate mineral that is rich in magnesium and iron and is commonly found in the upper parts of the mantle

Orocline—A bend or swing in the orientation of a mountain chain

P wave—A form of seismic vibration, generated by an earthquake, that travels through the Earth's interior—the motion of the vibrating rock is a bit like the vibration of a spring

Paleocurrent—Direction of transport of sediment, carried by water or the wind

Paleozoic—"time of old life"—period in Earth history between 250 and ca. 550 million years ago

Peneplain—Extensive plain or low relief landscape, formed after a period of erosion

Plates—The curved, rigid parts of the Earth's outer shell (synonymous with lithosphere) that move relative to each other

Plate tectonics—The behavior of plates

Postglacial rebound—The process by which the surface of the Earth returns to its original shape after being depressed by the weight of vast ice sheets

Precambrian—Portion of Earth history that predates the Paleozoic

Pyroxene—A silicate mineral commonly found in igneous rocks and the upper part of the Earth's mantle

Quartz—A silicate mineral composed entirely of silicon and oxygen

Radioactivity—Phenomenon exhibited by certain elements that are unstable and spontaneously decay to a daughter element—generates heat and occurs at a predictable rate that forms the basis for almost all modern methods of dating rocks

Rayleigh number—Parameter of a fluid that determines whether convection will occur

Red-bed(s)—Sandstone or interbedded sandstone, siltstone, and shale sequences that have a pervasive red color caused by the presence of minute quantities of iron oxide (hematite)—often river-lain sediments that have had a prolonged exposure to oxygen

S wave—A form of seismic vibration, generated by an earthquake, that travels through the Earth's interior—the motion of the vibrating rock is a bit like the sideways motion of a snake

Schist—A medium-grained metamorphic rock, usually containing platy minerals such as mica, that tend to break along numerous parallel planes

Sedimentary rock—Rock commonly made up of older rock fragments that were transported at the surface of the Earth, usually in water or by the wind, and then deposited in layers

Seismic section—Image of the outer part of the Earth, revealing the rock layers caused by seismic vibrations

Silica—Oxide of silicon (silicon dioxide)

Silicate—A type of mineral, which makes up most of the Earth's crust and mantle, composed largely of silicon and oxygen

Strata—The layers of sedimentary rock

Sub-Andes (Sub-Andean zone)—Eastern foothills of the Andes in Peru, Bolivia, and Argentina

Subduction (zone)—Process (or region of the Earth) by which the lithospheric plate sinks back into the Earth's mantle—associated with ocean trenches, earthquakes aligned in a Benioff zone, and arcuate chains of overlying volcanoes

Trench—Deep groove in ocean floor, about fifty kilometers wide and up to eleven kilometers deep, where the sea floor bends down and starts sinking back into the mantle in a subduction zone

Ultrabasic rock—Igneous rock with less than 45 percent by weight silica—mantle rocks such as peridotite are typical examples

Unconformity—The contact between two rock formations, which is the result of substantial erosion before the deposition of the younger formation

Viscosity—A measure of the runniness or stiffness of a fluid, measured in poises—viscosity of water is about a hundredth of a poise

Volcanic arc—Chain of volcanoes that lies above a subduction zone

Volcanic rock—Rock produced as a result of volcanic activity—usually formed when molten lava cools

Western Cordillera (volcanic arc)—Chain of volcanoes running along the western edge of the high Andes (Altiplano) in Peru and Chile, with individual volcanoes rising up to 2,000 meters above a general base level at about 4,000 meters above sea level

Further Reading

IN WRITING THIS BOOK I have drawn on numerous sources. The following is a short summary of both my own publications and those works that I have found particularly helpful. A well-illustrated account of many of the basic geological ideas covered in this book is given in *Earth Story—The Forces That Have Shaped Our Planet*, the book that accompanied the 1998 BBC television series of that name (Simon Lamb and David Sington, *Earth Story* [London: BBC Worldwide, 2003, and Princeton: Princeton University Press, 2003]). There is no better overview of the way processes at the surface of the Earth create the pages of the rock record, with all its myriad detail, than Arthur Holmes's classic *Principles of Physical Geology*, rev. 2nd ed. (Sunbury-on-Thames, UK: Thomas Nelson and Sons, 1977). The dictionary definitions at the beginning of each chapter are taken from the *Concise Oxford Dictionary*, 6th ed., ed. J. B. Sykes (Oxford: Oxford University Press, 1976).

CHAPTER 2

For a firsthand account of Peach and Horne's classic discoveries in the northwestern Highlands of Scotland, it is hard to find a better source than the original memoir of the Geological Survey of Great Britain (Benjamin Peach and John Horne, *Geological Structure of the North-West Highlands of Scotland* [Glasgow: His Majesty's Stationery Office, 1907]). Seminal works of early geologists or philosophers such as Leonardo da Vinci, Nicolaus Steno, Gottfried Leibnitz, Comte de Buffon, John Woodward, and James Hutton are compiled in K. F. Maher and S. L. Mason, *Source Book in Geology* (New York: McGraw-Hill, 1939). Sir Archibald Geikie also gives lively accounts of many of the famous stories connected with the early geologists in *Founders of Geology*, 2nd ed. (New York: Macmillan, 1905). The account of the flood story in the Ashurbanipal text is taken from Charles Officer and Jake Page's *Tales of the Earth* (New York: Oxford University Press, 1994). There are no more lucid accounts of early nineteenth-century thinking than Charles Lyell's *Principles of Geology*, first published in 1830 (London: Penguin Classics, 1997), and Charles Darwin's *The Voyage of the Beagle* (London: Penguin Classics, 1989). For some of the late nineteenth- and early twentieth-century debates about the nature of the Earth's crust, the significance of gravity measurements, and the early controversies about continental drift, Tony Watts (*Isostasy and Flexure of the Lithosphere* [Cambridge: Cambridge University Press, 2001]) and Naomi Oreskes (*The Rejection of Continental Drift* [New York: Oxford University Press, 1999]) provide entertaining and lucid accounts.

Chapter 5

For an account of the Spanish Conquest of South America and the famous battle on the banks of the Desaguadero, near Lake Titicaca in Bolivia, see John Hemming's wonderful book *The Conquest of the Incas* (London: Penguin Books, 1983). Arthur Holmes's *Principles of Physical Geology* (already cited) and standard first-year university textbooks such as Stephen Marshak's *Earth—Portrait of a Planet* (New York: W.W. Norton, 2001) provide a wealth of well-illustrated basic information about geological processes at the Earth's surface. For an account of our research on the Cenozoic rock record in the Bolivian Andes, with summaries of the relevant important stratigraphic and dating studies by other workers, see L. Kennan, S. H. Lamb, and C. Rundle, "K-Ar Dates from the Altiplano and Cordillera Oriental of Bolivia: Implications for the Cenozoic stratigraphy and tectonics," *Journal of South American Earth Sciences* 8(1995):163–86; S. H. Lamb et al., "Cenozoic Evolution of the Central Andes in Northern Chile and Bolivia," in *Orogens through Time*, ed. J.-P. Burg and M. Ford, Special Publication of the Geological Society of London 121(1997):237–64; S. H. Lamb and L. Hoke, "The Origin of the High Plateau in the Central Andes, Bolivia, South America," *Tectonics* 16:623–49; S. H. Lamb and L. Kennan, *The Andes of South America*, Oxford Companion to the Earth Sciences (Oxford: Oxford University Press, 1999). A reconstruction of the vast inland sea at the end of the Cretaceous that predated the Bolivian Andes is given by J. M. Rouchy et al., "The Central Palaeo-Andean Basin of Bolivia (Potosi Area) during the Late Cretaceous and Early Tertiary: Reconstruction of Ancient Saline Lakes Using Sedimentological, Paleoecological and Stable Isotope Records" *Palaeogeography, Palaeoclimatology, Palaeoecology* 105(1993):179–98. Michael Benjamin and co-workers' classic work on fission track dating in the Zongo Valley can be found in M. Benjamin, N. Johnson, and C. Naeser, "Recent Rapid Uplift in the Bolivian Andes: Evidence from Fission-Track Dating," *Geology* 15(1987):680–83.

Chapter 6

The main conclusions in this chapter are based on various original scientific papers. Bryan Isacks from Cornell University pioneered many important ideas about the origin of the Andes in "Uplift of the Central Andean Plateau and Bending of the Bolivian Orocline," *Journal of Geophysical Research* 86(1998):3211–31. For Susan Beck and colleagues' probing of the Andean crust, see S. Beck et al., "Crustal Thickness Variations in the Central Andes," *Geology* 24(1996):407–10. Estimates of shortening of the rock layers in the Bolivian Andes are given by B. Sheffels, "Lower Bound on the Amount of Crustal Shortening in the Central Bolivian Andes," *Geology* 18(1990):812–15; S. H. Lamb et al., "Cenozoic Evolution of the Central Andes in Bolivia and Northern Chile," *Special Publication of the Geological Society of London* 121(1997):237–64; J. Kley, and C. R. Monaldi, "Tectonic Shortening and Crustal Thickness in the Central Andes," *Geology* 26(1998):723–26. The case for the hypothetical "Calasaya nappe" is put forward in P. Baby et al., "Evidence for Major Shortening on the Eastern Edge of the Bolivian Altiplano: The Calasaya Nappe," *Tectonophysics* 205(1992):155–69. For an account of fossil floras collected in the high Bolivian Andes, see E. W. Berry, "The Fossil Flora of Potosi, Bolivia," *John Hopkins University Studies in Geology* 13(1939):1–67; K. M. Gregory-Wodzicki et al., "Climatic and Tectonic Implications of the Late Miocene Jakokkota Flora, Bolivian Altiplano," *Journal of South American Earth Sciences* 11(1998):533–60. The peneplain in the Eastern Cordillera of Bolivia is described by T. L. Gubbels, B. L. Isacks, and E. Farrar, "High-Level Surfaces, Plateau Uplift, and Foreland Basin Development, Bolivian Central Andes," *Geology* 21(1993):695–98; and by us in L. Kennan, S. Lamb, and L. Hoke, "High Altitude Peneplains in the Bolivian Andes: Evidence for Late Cenozoic Surface Uplift," in *Palaeosurfaces*, ed. M. Widdowson, Special Publication of the Geological Society of London 120(1997):307–24. For a simple account of the stiffness of the lithosphere, see A. Watts, *Isostasy and Flexure of the Lithosphere* (Cambridge: Cambridge University Press, 2001), and S. H. Lamb, "Is It All in the Crust?" *Nature* 420(2002):130–31. Detailed evidence for the stiffness and flexing of the South American lithosphere is described by H. Lyon Caen, P. Molnar, and G. Suarez, "Gravity Anomalies and Flexure of the Brazilian Shield beneath the Bolivian Andes," *Earth and Planetary Science Letters* 75(1985):81–92; D. Whitman, "Moho Geometry beneath the Eastern Margin of the Andes, Northwest Argentina and Its Implications for the Elastic Thickness of the Andean Foreland," *Journal of Geophysical Research* 99(1994):15277–89. Our results are published in A. Watts et al., "Lithospheric Flexure and Bending of the Central Andes," *Earth and Planetary Science Letters* 134(1995):9–21. Our students Stewart Johnson and Emily Black have extended this work as part of their theses. See: J. Steward and A. Watts, "Gravity Anomalies and Spatial Variations of Flexural Rigidity at Mountain Ranges," *Journal of Geophysical Research* 102(1997):5327–52, and E. Black, "Influence of Lithospheric Flexure on the Development of Mountain Belts, with Special Reference to the Central Andes," D.Phil. thesis, Oxford, 1999.

Chapter 7

For an overview of the magnetism in rocks and how it can be used in geological studies, see R. F. Butler, *Paleomagnetism: Magnetic Domains to Geologic Terranes* (Malden, MA: Blackwell Science, 1992). For a summary of Bruce MacFadden's and Pierrick Roperch's work on the rotation of blocks of crust in the Andes, using the magnetism in rocks, see B. J. MacFadden, F. Anaya, and C. Swisher III, "Neogene Paleomagnetism and Oroclinal Bending of the Central Andes of Bolivia," *Journal of Geophysical Research* 100(1995):8153–67; and P. Roperch et al., "Tectonic Rotations within the Bolivian Altiplano: Implications for the Geodynamic Evolution of the Central Andes during the Late Tertiary," *Journal of Geophysical Research* 105(2000):795–820. My own work is summarized in the following three papers: S. H. Lamb, "Vertical Axis Rotation in the Bolivian Orocline, South America, 1. Paleomagnetic Analysis of Cretaceous and Cenozoic Rocks," *Journal of Geophysical Research* 106(2001):26605–32; S. H. Lamb, "Vertical Axis Rotation in the Bolivian Orocline, South America, 2. Kinematic and Dynamical Implications," *Journal of Geophysical Research* 106(2001):26633–53; and S. H. Lamb and R. Randall, "Deriving Palaeomagnetic Poles from Independently Assessed Declination and Inclination Data; Implications for South American Poles since 120 Ma," *Geophysical Journal International* 46(2001): 349–70.

Many of the original papers that ushered in the plate tectonics revolution are reprinted in *Plate Tectonics and Geomagnetic Reversals* (San Francisco: W. H. Freeman, 1973).

Chapter 8

See *Earth Story* (already cited) for an illustrated account of Harold Wellman's discoveries in South Island, New Zealand. Wellman's prescient 1955 paper on active tectonics in New Zealand was published in *Geologische Rundschau* 43:248–57. Hubbert and Rubey's classic 1959 work on the lubrication of faults by the high pressure of water was published in the *Bulletin of the Geological Society of America* 70:115–66, 167–205. Dan McKenzie's 1967 work on plate tectonics on a sphere (together with Bob Parker) was published in *Nature* 224:125–33. In 1972 he described the active motions of blocks of crust in the eastern Mediterranean (*Geophysical Journal of the Royal Astronomical Society* 30:109–72). Peter Molnar and Paul Tapponnier's classic 1975 and 1976 papers on the major faults in Asia were published in *Science* 189:419–26 and *Nature* 264:319–24. Dan McKenzie, working together with Philip England and James Jackson, produced two seminal papers on the fluid theory of continental deformation: P. C. England and D. P. McKenzie, "A Thin Viscous Sheet Model for Continental Deformation," *Geophysical Journal of the Royal Astronomical Society* 70(1982):295–321; and D. P. McKenzie and J. A. Jackson, "The Relationship between Strain Rates, Crustal Thickening, Paleomagnetism, Finite Strain and Fault Movements within a Deforming Zone," *Earth and Planetary Science Letters* 65(1983):182–202. I subsequently expanded on the conclusions of the McKenzie and Jackson paper (see Lamb, "A Model for Tectonic Rotations about a Vertical Axis," *Earth and Planetary Science Letters* 84[1987]:75–86; and Lamb, "The Behavior of the Brittle Crust in Wide Zones of Deformation," *Journal of Geophysical Research* 99[1994]:4457–83). Dick Walcott's important 1984 paper on the flow of New Zealand, using information from nineteenth-century New Zealand surveys, was published in the *Geophysical Journal of the Royal Astronomical Society* 79:613–33. Edmundo Norabuena and colleagues' GPS measurements of ground movements in the Peruvian and Bolivian Andes were published in *Science* 279:358. For my own work on the flow of the Bolivian Andes, see Lamb, "Active Deformation in the Bolivian Andes, South America," *Journal of Geophysical Research* 105(2001):25627–53; and Lamb, "Vertical Axis Rotation in the Bolivian Orocline, South America, 2. Kinematic and dynamical implications, *Journal of Geophysical Research* 106(2001):26633–53.

Chapter 9

A good overview of the origin of igneous rocks, the continental crust, and many other issues connected with the evolution of our planet is Minoru Ozima's *Global Evolution of the Earth* (Berlin: Springer-Verlag, 1987), and *Earth Story* (already cited). Arthur Homes's *Principles of Physical Geology* (already cited) is always worth looking at for clear and concise explanations. For our work on helium in the Andes, see L. Hoke et al., "The Evidence for a Wide Zone of Active Mantle Melting beneath the Central Andes," *Earth and Planetary Science Letters* 128(1994):341–55.

Chapter 10

This chapter contains rather more speculative ideas than have hitherto been discussed in this book. Some of these I have developed while writing the book (especially those on the role of sediments and climate in the rise of the Andes) and are only now being published in scientific journals (see S. H. Lamb and P. Davis, "Cenozoic Climate Change as a Possible Cause for the Rise of the Andes," *Nature* 425[2003]:792–97). However, for good general overviews of the many interconnections among mountain building, global climate, and the evolution of life, see *Earth Story* (already cited); Open University second-level science course book, S269, *Earth and Life—The Dynamic Earth* (Milton Keynes: Open University, 1997); *The Book of Life*, ed. Stephen Jay Gould (London: Ebury, Hutchinson, 1993). Maureen Raymo and William Ruddiman's ideas about the link between the cooling of the climate and the rise of mountain ranges can be found in M. E. Raymo and W. F. Ruddiman, "Tectonic Forcing of Late Cenozoic Climate," *Nature* 359(1992):117–22. For our work on helium in Tibet, see L. Hoke et al., "The Southern Limit of Recent Mantle Melting in Southern Tibet, Constrained by Helium Isotopes," *Earth and Planetary Science Letters* 180(2000):297–308. For an account of theories about the uplift, rifting, and extension in Tibet, see P. C. England and G. A. Houseman, "Extension during Continental Convergence, with Application to the Tibetan Plateau," *Journal of Geophysical Research* 94(1989):17561–79; P. Molnar, P. England, and J. Martinod, "Mantle Dynamics, Uplift of the Tibetan Plateau, and the Indian Monsoon," *Reviews in Geophysics* 31(1993):357–96.

Index

Italicized page references refer to illustration pages